# GET REAL, PEOPLE!
## How to Use Social Media for
## REAL-ationship Marketing©

*by Shelley R. Roth*

SPRINGBOARD PUBLISHING

SPRINGBOARD PUBLISHING

www.springboardworks.com

*Jacket design by Shelley Roth and Energy Arts Alliance. Edited by Energy Arts Alliance.*

LIBRARY OF CONGRESS CATALOGING-IN-PUBLICATION DATA
Roth, Shelley R.
GET REAL, PEOPLE! *How to Use Social Media for REAL-ationship Marketing*© / Shelley R. Roth. - 1st ed.
p. cm.
1. Marketing & Sales 2. Business & Economics
3. Social media

ISBN: 10-0983870403
ISBN: 13-9780983870401

PRINTED IN THE UNITED STATES OF AMERICA

First Edition

*Sign up for a free e-copy of "Engaging Customers with Social Media". Visit: on.fb.me/engagecustomers.*

# CONTENTS

Introduction
Welcome to *GET REAL, PEOPLE! How to Use Social Media for REAL-ationship Marketing*©

Chapter 1
From Software XYZ to transparency, OR How I got real and wrote this book

Chapter 2
The road to getting real & interpreting realness online

Chapter 3
What is real (online)?

Chapter 4
Lessons from nature: Being real is natural

Chapter 5
Approachability is the new handshake, trust is the new currency

Chapter 6
Getting real: Dropping the hidden agendas

Chapter 7
Reading and perceiving digital body language

Chapter 8
The nuts and bolts of getting real online: Specific ways of using social media effectively

Chapter 9
REAL-O-METER©: How you communicate online and ensuring your words reflect the real you

Chapter 10
Beyond the digital frontier: Getting real in the real world

Acknowledgements

Appendix I
Master list of all links in the book

Appendix II
Interview Survey Results

INTRODUCTION

Welcome to *GET REAL, PEOPLE!: How to Use Social Media for REAL-ationship Marketing*©.

I am a recovering sales shark. I spent a lot of time being someone I wasn't only to discover my true passion and identity later in life. Do you relate to that, or know someone like that? Then this book is for you!

The goal of this book is to let us know it's okay to be REAL and to open up to who we really are being in life, with a focus on being real online. Being real is simple, but NOT easy. It's about all of the beliefs and assumptions we created for ourselves over the course of our childhood and early adult life that define, or that WE THINK define, who we are. Our perceptions rule our world...and our perceptions are what we choose them to be! PERCEPTION = REALITY. Everything in life is a choice, and we can always make new choices.

Here's a great simple example of how perceptions can differ: Two footwear marketing scouts are sent to a remote region in Africa to determine if there is an opportunity to expand business there. One sends back a report stating: "Situation hopeless; no one wears shoes here." The other states: "Wonderful business opportunity; no one wears shoes here."

This simple example shows how our interpretations and attitudes either propel us or stop us.

You are born real. This book supports realness with awareness and tools to express and detect truthfulness and authenticity. It supports the concept of "giving to grow".

A big part of detecting digital truthfulness is being aware and alert that there are people online who may be wearing masks and not being authentic.

*GET REAL, PEOPLE!* is about helping us be comfortable in our own skin, to live our purpose and passion and to listen to our instincts, higher self, gut, heart - whatever you are comfortable calling it.

*GET REAL, PEOPLE!* raises thought-provoking questions to help you be aware of whether you and others are being real online and in life.

At the time of the first publishing of this book, we're developing a fun and intriguing realness tool and app, the REAL-O-METER©. Purely for entertainment purposes, REAL-O-METER© makes an experimental attempt at measuring how real or unreal social media posts may be. We are trying to do this by detecting the presence and frequency of positive and negative key words used in social media posts, based on the opinion of this author and the editors. We plan on updating the measurement criteria often, so you can use the tool over time to measure the progress of your "realness quotient."

**CHAPTER 1**
**From Software XYZ to transparency, OR**
**How I got real and wrote this book**

So, here's my story on NOT being real for the first two-thirds of my life...Yup, two-thirds. Wow, what a waste of time and energy being someone I wasn't. Why was I not ME? I truly believed that the "real me" wasn't like "normal" people. I particularly didn't see "rich" and successful people like the real me. Both personally and professionally, I had on every mask I could use to be anything but ME! Why?

Let's start from the very beginning when I was labeled a "tomboy". I was ridiculed (and not just by school-age friends but also by many adults in my life) for wanting to excel at sports, be competitive, win at everything I attempted. Competition was not "ladylike"...If I had a dollar for every time I heard that term, I could feed all of Africa!

I mean, back then, ladies and little girls were supposed to behave a certain way. Dainty? Not me! Quiet and unassuming? Not this one! Speak when spoken to? Forget about it....Nope, I didn't fit the mold of whatever someone (was it Dr. Spock?) decided good little girls were supposed to be: sugar and spice and everything nice....Yeah, right!

The other thing I heard repeatedly during my childhood and teen years was, "You are a miserable soul!" In my experience, perception is reality, so hearing this over and over, well, I naturally thought I was basically a miserable soul.

Now this is a book about business and being REAL online and being aware of digital body language, so I will not digress too much to my childhood. I just want to set the stage for my great awakening: **the fact that being REAL is a gift.**

**My life's purpose is to encourage others to be real by being real myself.** I'm here to help remind people that being REAL is what we all should strive for – no masks, no pretense, just authenticity with the by-product of joy.

So professionally, where did I start? Well, I was always a businessperson – from my early years with very successful lemonade stands, ice ball stands, magazine sales and selling cutlery door-to-door. All the ice ball stands, lemonade stands, Girl Scout cookies, magazine sales...I mean, I was THE salesperson! It was all about excelling and winning. I watched my father sell, and he was very successful, winning trips to many first-class exotic destinations as a result of selling. (Actually, he was selling snow to Eskimos, and again, the learned behavior was that it was okay to do this.)

But my life as a sales shark was not to be, at least not yet. When I was young, "businessperson" just didn't fit the girls' career options when it was time to pick a college major. So I started off going the safe route, the route every girl opted for when I was graduating: nurse, secretary or teacher. Teacher was the one of the three that felt most appropriate. (I was always told I was a Pied Piper with children.) And in retrospect, it was a great foundation for who I am today, albeit, most likely not the perfect route for this "tomboy" (or so I thought).

So, teacher was a safe place for a girl who felt like a boy. It was a noble career, and I totally enjoyed working with elementary-age students. I spent eight years in education as a teacher, guidance counselor and assistant principal – all enjoyable, all successful, but no brass ring. I wanted "stuff," a lot of stuff, stuff = success, right? So I thought I would follow in my dad's footsteps. I sold my house, quit my job, cut my hippie hair and moved on down with a U-Haul to Houston, Texas.

Yep, sales would bring me all the things I thought would satisfy this lost, "miserable" soul.

Sales was a place for elephant hunters: "Kill or be killed," "scorch the earth," "take no prisoners"... and I never felt that way. I was a farmer, a nurturer. I didn't want to eat what I killed. I didn't want to kill; I wanted to nurture. But that didn't matter. Part of me wanted those trips to Hawaii, those luxury cars, the big house. At least I thought I did, and that part of me won out over the nurturer.

My dad was motivated by stuff, I thought. (I never really got to ask him if that was really true.) I think the middle class dream is what gets to many of us, and we think success is the stuff we acquire. We think it's a symbol of who we are and what we have accomplished. But, I digress...

So, I was rewarded for the things that I ultimately learned I truly didn't believe in, things I didn't think were really signs of success, things that never really brought me happiness or joy.

Before I was "discovered" by a top computer software company, I paid my dues during my first two years in Houston. From selling frozen food plans (mostly chicken because the bones weighed a lot) to people on welfare to sun-tanning salon franchises, I truly went through the school of hard knocks sales. Before being discovered by that computer company, I sold the regional manager a swimming pool and, alas, I was plucked from obscurity into a top-notch company.

That first Fortune 500 I worked for had 80 salespeople, and I was the number one quota achiever in my first year. I "won" a two-week, all-expenses paid trip to Tahiti, and I had 79 male counterparts mad as hornets about that! Being the practical person I am, I decided to forgo the trip to Tahiti, and instead took a first-class trip to Hawaii for a week, using the balance of the award to purchase many items that I had never owned before: patio furniture, huge TV, fancy lamps, etc. Wow, that was some first year with the big company! But what did I want? What were my targets, my monetary rewards? Cars, houses, boats??? Nope. When asked by the sales manager to list all of the "things" we aspired to and wanted (which was used to motivate us to sell more), peace, harmony, joy were at the top of my list...Talk about a fish out of water...But boy did I learn to play the role, to wear the mask!

With all the past sales experiences that I've already shared, there was something else that really taught me to be a stud salesperson. It happened when I was five years old. It was around Halloween, and we were at a costume contest at the neighborhood community center. I was dressed in a homemade bunny costume and really feeling it. Hip-hopping my little heart out. And then, I WON the contest. I was called up to the stage to accept my reward and walked right up bold as could be. The Master of Ceremonies was announcing this into his wired microphone, as he handed little bunny Shelley a shiny silver dollar and...BAM!...The little bunny literally had the shock of her little life...The microphone wasn't grounded, the coin acted as a conductor and - WHAM! - an electric shock was transmitted from the mic through the coin into my body. What's a bunny to do?

This was the moment the salesperson was born. Did I listen to my heart and cry and ask for help or comfort? Did I yell out for my mom to come up there and take me away? NOOO, I stood taller, stronger and pretended that there was nothing wrong. That was the moment I listened to my head, and it said, "Don't be a sissy. Don't cry. Don't show emotion. PRETEND there's nothing wrong. PRETEND you are happy. You don't want anyone to see your real feelings or heart... It's all about looking good!"

From that day forward I was terrified of public speaking, but oh, what a great actor I became!

After 20 years being the great salesperson, working for Fortune 500 companies, listening to my managers tell me how to close more deals and "winning" trips to Hawaii (6 times), Bermuda, Bahamas, Lake Tahoe, Cancun, Amsterdam, Belgium, China, buying the house, the fancy cars, all the stuff, well, I still didn't feel fulfilled. What was my purpose? Why was I on the Earth?

Those questions must have set something in motion. After all those years at the top, I was fired because I finally couldn't do what I was told since in my heart I knew selling ice to Eskimos wasn't a noble deed.

After being kicked out of the corporate nest, I started my own business and have worked with companies over the last decade on what I know, growing sales. A couple of years back, I was intrigued by social media, so, sponge that I am for learning, I decided to create curricula on various social media tools and combine my passion for business with my love of training, and here I am today. The more I worked in social media, the more I realized that we all need to be aware of what is REAL out there. That's when I was led by my wonderful energy advisors to discover my reason for being on the Earth and to encourage others to do the same: to be REAL. And this was the one thing I never had confidence in being. This was the one thing I thought was BAD.

Every time I just WAS from my heart, it got slapped down... But not anymore.

REAL is good. REAL is king. REAL is what we all should strive for. REAL is the gift I was given.

**So now, I declare my vision and purpose: To help and encourage people to be REAL. To listen to your HEART. It knows what's right.**

### MORE OF THE STORY: SOFTWARE XYZ

From the minute I began the intensive interview process at the Fortune 500 software company to the day I departed, the experience never felt right to me in my heart. It felt great for my ego; I mean, who wouldn't want to work at Software XYZ? People were and still are clamoring to work there. They have workout facilities, a 4-star restaurant, hair and nail salon, banking and basketball court – all on the premises! This was the bait on the hook along with a fantastic beautiful first-class building with top-notch furnishings and shiny everything everywhere. Oh, and the salesperson was king for sure.

From the moment I parked my vehicle in the largest parking spaces I have ever seen in a garage (didn't want to scratch those BMWs, Mercedes, Porsches) to the moment I left that place in the evening, my stomach hurt. I never felt real there. I had a mask on the entire time.

I played the game, and I played it really well. I learned to act the perfect salesperson, yet, I am sure in retrospect that anyone who cared enough knew I was a fish out of water. I really didn't want to play this game, although the incredible financial and intrinsic rewards were enough to keep me there for a couple of years. Until I got laid off. Fired might be more appropriate. I was just beginning to question my instincts, and I wasn't just rolling over any more. I wasn't performing like the trained seal I had been for so many years as the typical sales shark. I was questioning my managers, questioning my heart. And, ultimately, I was let go.

Now, this was an absolute shocker for me. It was right before Thanksgiving, and I went home shell-shocked. I woke up in the middle of the night basically crying my head off in the front of the house so as not to disturb anyone else. I thought I had lost my identity, who I was, the big-shot salesperson.

What now? I had to be with my entire family in two days to celebrate Thanksgiving. (Little did I realize then that down the road a stretch I would come to understand that this truly was a reason to give thanks!) So as I was bawling my self-pitying head off, I heard a cat crying. "Meow, meoooowwww" - a cat in distress. I forgot "me" for a minute and listened and sure enough, it sounded as if a cat was in the wall of the dining room. Long story short, my cat had come through the breezeway and fallen through the wall. I won't get into what it took to rescue that cat - which we did - but what a gift that was!

It got me out of myself, made me realize - if only briefly, but it was a glimmer! - what was important in my life, and the net of it was I am NOT a sales shark. I am not a hired gun who wants to take no prisoners.

I wanted to just be my real, authentic self and listen to my heart. Listen to what was real and not question that THIS was who I truly am. Be confident in myself, my true self.

So, if this book serves a purpose to help all the recovering sales sharks LISTEN to their true selves, then I am happy.

**TRANSPARENCY**

I would never say all those years in sales not listening to my higher self, my instinct, my spirit and my soul were wasted because I believe that we are all here for a purpose, and I needed to learn those lessons to be able to share my experiences with transparency.

I was always real, honest, transparent, and I allowed myself to "hear" that this was not the way to be. Being real was just not in vogue. I'm not sure it's any more in vogue now, but I do know this: If we all embrace who we are and "be", just be who we are, well, the world would be better served.

Even when I was doing very well in sales as a sales shark and using my gift of communicating with all people, it was always about relationship-building for me. I have always excelled at that. 80% of my work during my sales career was starting and building relationships with business partners and resellers around our products and services. This spoke to me! Being social was what I was about. Listening, trusting and being transparent garnered many a client and friend.

It's the sales quotas that got in the way. The fear of lack. The "coming from scarcity" vs. coming from abundance that made me fear-based. Just recently, I met a salesperson who was just so sunny and open and real, and I wondered, hmmmm, this person must not be a salesperson because they just come from abundance, as though their "quota" was made and they could just be the wonderful person they were.

I came to find out that the company they work for doesn't believe in quotas. They want their reps to represent the company without the pressure of a quota. The difference it makes in this "sales" rep's relationships is astounding! Everyone wants to be/talk/think and sit with this rep who comes from abundance vs. need.

I was so motivated, I now include a story on this company in every presentation and training class that I do. The company is the largest email marketing company on the planet.

So, all of this transparency has led me to this book on digital body language and being real online. This book's goal is to bring an awareness to who we are, to be real, to listen to that inner voice.

My energy advisors believe there are a lot of sales sharks out there just waiting to shed that shiny sales skin and be the true, authentic people they were born to be. So, let's say this chapter is dedicated to RECOVERING SALES SHARKS throughout the world! I mean, I was one, or at least I tried to convince myself I was!

Footnote: Software XYZ is a wonderful company that is very generous, makes fabulous software and is right for a lot of people. It just wasn't right for me.

## CHAPTER 2
### The road to getting real & interpreting realness online

Three and half years ago, I was changing direction in my company and wanted to learn all I could on social media so I could develop curricula and train others on integrating these social tools into their marketing plans. I was doing research on who might be good mentors for me around social media usage.

I found a guy with a video. He had thousands and thousands of Fans. Needless to say, newbie that I was, I was totally impressed by the sheer number of Fans this guy had. A common misperception in social media is that number of Fans or followers or connections equals a better product or service.

So, this guy's video was engaging, and I liked what I heard. He seemed smart and was VERY confident, however, I was wanting to do a bit of due diligence, so I started Googling him, drilling down on him. I found, after chasing down many rabbit holes, that the guy was a convicted FELON!! still living in the town where he was convicted! (That's not to say that felons can't overcome their past and be very positive citizens contributing to their community.)

Needless to say, I stepped back from the computer. It gave me a real JOLT, an awakening, an aha moment. I started thinking, how do you read people's digital body language and the digital footprint they leave out there in digital land? And how do you know or at least have an idea of who you want to deal with, who you want to listen to, who you want to buy from, who you can trust, who is authentic, transparent and REAL?

It started me thinking that maybe there is a real science/art to identifying how to read people online. How to read their: integrity, values, honesty, spiritual basis, do they really care vs. act like they care? A picture is worth a thousand words, so they say, and photos online are the quickest way to "read" someone. You immediately feel drawn to them or not. Not just the smile or lack of a smile, or what they are doing in the picture, but actually how the picture's energy resonates with and is interpreted by the viewer. It has to do with digital body language/digital footprint or the energy the person was exuding when that picture was taken. That's one of the first areas you want to consider when you post that picture of yourself. Is it inviting? Are you smiling? You want to look engaging, well meaning. And frankly if you ARE engaging and well-meaning and true and authentic and transparent, this will show through that picture naturally. It's our own personal BAROMETER that needs to be fine-tuned to recognize how to interpret online impressions. That's why we've included a checklist of words we're using to develop the REAL-O-METER© app with this book.

The REAL-O-METER© app will use positive and negative keywords to try to identify how real you and others are being in their social media posts. We selected a range of keywords from potentially high authenticity (e.g., caring, common good, goodwill) to potentially low authenticity (e.g., one time only, today only, everything must go). It's a fun way – for entertainment purposes only – of experimenting with whether we can determine if a blog, Facebook page, LinkedIn profile and other electronic documents that can be scanned, contain words we deem as "more authentic" or "less authentic". We used our own reading of many, many profiles, blogs, etc. to establish this subjective list of keywords and their relative weighting as more or less authentic.

**The basics for reading people's digital body language are simple:**

Let's start with the **picture** that represents you:

1. When you put your picture out there, that's the FIRST connection people will make to you. You want to assure you have an energy vibe that draws people into you. What is an energy vibe? How do you determine that? It really is a gut feeling you get when you are live and in person. We have done survey research to see what our focus group "reads" from pictures and will

share that with you in Chapter 7 on reading and perceiving digital body language.

2. What you **write/post** and contribute can tell lots about you. Are you giving what people want to hear? Are you engaging them? I mean truly engaging vs. what's in it for you? <u>Engagement</u> is the most important way to build your community using social media. It basically takes the form of communicating what matters to the members of your community. Whether you are replying to a blog post, posting a status update to your Fans on Facebook and then responding to their comments and queries, participating in group discussions on LinkedIn or tweeting what your followers are wanting to hear, engagement is the secret sauce online.

   Are you asking for what you want or what <u>they</u> want? Are you giving to get vs. giving to grow? Are you making it easy for your community to respond to you? How are you being perceived? Do you make it seamless to learn about "brand you"? Do you make it easy to access information, or hard?

3. When you are **sharing information**, is it of value? What impression are you leaving? Are

you about you or about others? Is it spam? One of my red flags on digital impression is inside of groups. The spirit of groups, whether on Facebook or LinkedIn, is to share and add value to the group. Now that means not promoting your products and services unless they are relevant to the topic at hand. I have left many a group when the intention of the group administrator becomes to use that platform for selling themselves or their products. Most social media tools now have a way of self-policing this issue so the true spirit of the group can occur. As an example, in group discussions, groups don't want to hear about my workshops coming up. I don't want to share my workshops. It is tempting, but it's not the digital impression that I want to portray out there. It's important to always think before sending anything online. Always be asking yourself, "What's in it for them?", not "What's in it for me?"

Now IF the discussion is specific to what you have to offer, well by all means, jump in and share what you can offer, whether it's your product or service or someone you recommend. The key is IF it's **relevant** to the conversation and it adds value.

## WHAT WE CAN READ FROM PICTURES AND WORDS

There is a whole science to buyer's behavior when they go to a company's website. Eloqua is a company that has developed software that monitors buyers' behavior on a website to determine where a prospect is in their buying cycle based on actions they take on a business website, such as downloading brochures. They look at "digital body language" or patterns of actions of the prospective buyer and assist in reading that buyer's behavior.

The digital body language or digital impression I am talking about in this book is about how people reading your blog, posting on your Facebook wall, responding to your LinkedIn group and interfacing with electronic communication in general sense  you and what you are portraying when you interact or post online. Creating community and wanting to have your followers, fans and connections stick with you and listen to what you have to say is the objective of being real online and truly communicating who you are and what you stand for. It's about "REAL-ationship Marketing©" – the new "get real marketing" via social media. Why? Because you are being authentic, trustworthy, just plain REAL, AND you are providing information that is valuable to them. WIIFT? (What's in it for them?)

From the vantage point of building a community of customers, prospects, colleagues and partners who are utilizing social media tools, digital body language is a crucial piece that needs to be understood. **My intention with digital body language and this book is to make people AWARE of how they may be interpreted online through all aspects of their online persona.** So, if I can truly trust that my realness online will bring more rewards than I can imagine, then the goal is NOT to be selling anything. It will always be about giving to my colleagues, prospects, partners and clients while knowing that when they need what I have to offer, they will then engage my professional services. And I will never have to toot my own horn, "take no prisoners" or close a deal because it's the end of the quarter and numbers must be made.

Let's look at some examples of the obvious and easiest way to read digital body language, the picture you post on Facebook, LinkedIn and other social sites that require this.

Here are some pictures to look at.

Provocative...Not such a good idea to reach your business audience

*Photograph: Louisa Stokes, freedigitalphotos.net*

Angry and threatening...
Is that how you want
people to see you?

*Photograph: Ian Kahn,*
*freedigitalphotos.net*

Happy and light-filled...
That's more like it!

*Photograph: Dundee Photographics,*
*freedigitalphotos.net*

Partying...Remember HR and
college admissions offices
research prospective
employees and students
via all social media!

*Photograph: photostock,*
*freedigitalphotos.net*

Ask yourself: What do you get from the pictures shown?
What engages you? Who do you resonate with? And

why do you feel drawn to them or not drawn to them? What would you do differently? Why?

The next step in "reading" someone is via the written word. If we didn't have the benefit of seeing pictures, I believe that with a lot of practice, we could read people's bios, blogs and status updates and determine when someone is trustworthy, giving, forthright and honest vs. selfish, self-promoting, self-centered…My advisors at Energy Arts Alliance (**energyartsalliance.com**) can already do this really well and help others learn how to do it, too. They created Digital Muscle-Builder Exercises for this book to help you hone these skills yourself.

As I've mentioned, there is a science around digital body language and what you can interpret once you are in tune with some of the signals out there in the online world. If you don't have a picture posted on LinkedIn or Facebook, **many people will not even consider responding to you. Be sensitive to how you may be interpreted and what the lack of a picture may say about you.**

Personally, I tend to interpret the absence of a photo as meaning that you weren't thoughtful enough to learn how to upload a picture or get a picture that represents you prior to publishing your page or bio. On the other hand, the real reason you don't have a photo up may be that you're concerned about privacy issues due to past history and don't want to be exposed online with a photo. If that's the case, then maybe social media isn't for you.

The presence or lack of a photo is an indicator of digital body language that speaks volumes about a person without saying a word – or maybe causes you to be "saying" something you didn't intend based on someone else's interpretation! Without in-person cues, the potential for these variations in interpretation becomes greater and the gaps between intention and perception can get wider, leading to potential misunderstanding and unintended consequences.

**DIFFERENT LEVELS OF DIGITAL FOOTPRINTS**

What are some other indicators of digital body language and the impression you leave for others to interpret? Well, there are different levels of the digital footprint you leave: An example I like to use is the "FREE something". Now free is not bad; as a matter of fact, I give a free document to anyone who signs up to become a Fan of Springboard Works. So, free is good. The point is to just be aware of the FREE offer online. FREE webinars, FREE workshops, FREE books.

Just ask yourself what do they want, and what will I be receiving once I accept this free offer? Usually free is a loss leader with the goal of selling something else to the attendees or respondents. An advertising agency that offers a FREE session on social media may be wanting to acquire clients that need PR work or publications produced. Other obvious self-promoters may be chest thumpers. They are all about them, and it's very apparent they are just about selling what they have and promoting who they are. Most people don't want or care about what you have or do UNLESS there is value for them. UNLESS they have a need at that time for what you have to offer. So always come from making deposits in their "bank" vs. asking for withdrawals. When they need or want what you have, they will come. They will know what you do and what you have to offer of value because you "give to grow".

Ask yourself the question, "WHATS IN IT FOR THEM [WIIFT]?" all the time. If you can answer this question before you post anything, and if you can truly answer that there is a value for them, that's the goal. GIVE TO GROW! I learned this "give to grow" philosophy from Energy Arts Alliance, editors of this book, contributors and my advisors. What I realized in using "give to get" was that this attitude was falling short of truly giving in the spirit of NOT receiving anything in return. The "give to grow" mentality allows an opportunity to practice and exercise your digital body language muscle while truly NOT thinking about "what's in it for me?" Giving to grow is a practice and must be repeated both online and in real life. What you put out will come back. It's the law of physics that for every action, there is an equal and opposite reaction.

Think about it. If you gave, meaning helped those that needed help and asked those you encounter, "What are you working on, and how can I help you?", then your business would soar. There would be no agenda, no sales quota, no external manipulators. There would be no reason NOT to be yourself at all times. There is NEVER a good reason not to be yourself. Yourself is your highest self.

We often go beyond not being ourselves to MINIMIZING ourselves for whatever reasons we have, to make others look bigger, smarter, stronger and because we find it easier to give vs. receive. (Nothing wrong with that, however, we must swing the other way as well. We do have bills to pay!)

Why do we do this? I do this to make others look bigger. Why? Because all my life I had it made up that I wasn't smart enough. That's right, with two Master's degrees, I wasn't smart enough, and my mind talk held me back from being who I was for fear of being judged as not smart enough. Your mind talk might be that you are not pretty enough, successful enough....whatever your "not enough" is, it's all made up!!!

If we are going to make it up, why not make it up positively? That's the other side of the continuum of not being real. On the one side, you might be more comfortable giving vs. receiving. On the other side of the spectrum, there is the BOASTFUL self - me, me, me. Listen to me, I am the best, I am the most important, I know everything. What does this look like? Take, take take, blowhard, over-projecting voice, over-projecting actions, wanting all the attention...Hmmm, looks like I relate to this as well.

Both sides of the spectrum are ego-based and false and not from the heart. The heart is pure and forgiving and loving and has no need of boasting or minimizing OR shrinking to be less than. The REAL self, the higher self, is in the middle of the spectrum. Not boastful, not bashful, just being. Being present, no agendas, no outcomes, no future, no past. JUST PRESENT.

As pioneers in the field of digital body language, all of us can translate the "real" person we are online into our everyday world. In fact, it's crucial that we are heartful people all of the time. Our healthy existence depends on it!

**DIGGING DEEPER**

We've explored the power of the online photo and the impact of words. But the third layer is the hardest to read: The reality of who is truly giving from their heart vs. the wolf in sheep's clothing. It is sometimes difficult to know if someone has a hidden agenda or is truly living the philosophy of "giving to grow".

For example, one of my Facebook gurus spends a lot of time answering any questions on her wall that Fans might ask. I thought she was an example to me of how to use this philosophy. But as I have continued to follow her, I learned that she only <u>appeared</u> to come from this "give to grow" way of being. I found out from her own statements that she always has a marketing agenda behind her giving. Needless to say, I was disappointed to learn this; however, in the spirit of giving, I've learned a lot from her and now can live the philosophy myself in the way I support my Fans on the Springboard Facebook business page (**www.fb.com/springboardworks**).

Now please don't get me wrong: As I mentioned before, everyone needs to make money and has bills to pay, so there is nothing wrong with making money. It's just a new way to think about selling, or "non-selling" as I call it. You don't have to be a sales shark. You just have to give value to people by loving to discover what they need and loving what you do. They will know who you are and what you have to offer, and you will NOT have to tell them over and over.

So, a rule to live by - not just online but every day - is to come from abundance vs. scarcity. This means the glass is always more than half full, that you have what you need. Your digital body language will reflect this. Practice makes perfect. Everything in life is a choice. Choose ABUNDANCE!

I believe if you are doing something you are passionate about and love, and if you don't have fear or come from scarcity, your digital body language will be true and clear, and the energy around your impression will be interpreted positively. It's crucial when you are out there as a prospective provider and receiver of services to be able to read and be read in a positive way.

Interestingly enough, there is another book out there on digital body language as it relates to buyer behavior. Companies use the author's software to interpret buying habits. It's important to be aware that Facebook with the advent of the LIKE and the SEND buttons is taking this even a step further. Knowing what you're friends LIKE and are interested in is more powerful than ever for advertisers. Now they can utilize what BRANDS you share in common. That is some kind of powerful digital body language to help them pinpoint, advertise and establish which groups of your friends or colleagues LIKE a certain brand or service! Herd mentality, here we come!

The digital impression created by shifting your thinking around your wants to What's In It For Them? (WIIFT) is a huge shift in how we do business with people. With the advent of social media and the awareness of digital body language, the sales sharks are going the way of the dinosaur. Buyers are a lot more sophisticated and social media has given them not only a voice and an outlet for expressing their likes, dislikes and views, but also a means to review and research what they are interested in anonymously.

For example, you can lurk in various Facebook or LinkedIn groups without revealing that you're there or who you are. And you can post comments on blogs anonymously. With Google Alerts, you can be alerted any time something is being said about something you're interested in without anyone knowing you are receiving that information.

We as human beings can be very egocentric and motivated by "getting" intrinsic rewards. When you are marketing using digital media, think about whether you are giving to get. Think about your target market and what they are motivated by. What is going to motivate them to action? There is a way to do this that shows the footprint you are leaving and the impression you are making truly is about you wanting to assist and add value in the digital world. It takes a bit of adjusting your thinking, but you'll get there. Test the waters, see what people respond to. Monitor activity through tools like Google Analytics (**www.google.com/analytics**).

Facebook, LinkedIn, Twitter and YouTube... All of these top social media tools are all about sharing information and therefore a platform and a means of creating your digital body language impression or footprint.

Business leaders need to do the right thing with social media and reap the reward... CONSCIOUS businesses that truly care about making a difference and strive for **EXTREME** customer service will be noticed and recognized. **Putting the customer first and listening to what they are saying is crucial.** Comcast is a great example of going from horrible customer service to turning it around using social media as their platform for extreme customer service.

Comcast was notorious for not the best service. They turned it around by listening to their customers via Facebook and Twitter. They have employees who are tasked with "listening" to what is being said and then responding in a timely fashion. So, for example, someone tweets that the Comcast technician didn't show up for a technical call to fix their cable TV. Comcast monitors the negative statement and uses it to turn the situation into a positive experience for the unhappy customer with all of the Twitter community watching what is being "said" and done in response to the challenge.

Another example of this is a large pizza chain featured in a video that went viral and put the chain in not such a good light. Instead of running for the hills and burying their heads in the sand, the pizza chain used the viral video as a means of apologizing via tweets and comments to their many followers and Fans. They not only made lemonade from lemons, but ended up turning legions of negative comments into positive comments about how great they did "listening" and making a difference with their customers. Other companies now strive to provide this level of care and service. It all comes back to GIVING to GROW! And the impression you leave online.

## THE CHANGING MARKET LANDSCAPE

What's changed out there? Why this amazing shift in social consciousness? Is it that we Baby Boomers have realized what is truly important in our lives? Is it Generation X and Y not settling for being sold a bill of goods? I'm not sure of the answer yet, but I am sure that social media is contributing to that shift in a big way.

When we are online, connecting, I think most people can "sniff out a rat", and they know when you are self-promoting and when you only care about what's in it for you. I believe if we are in tune with ourselves and others, we can also tell when people are doing things for the right reasons. Truly caring about the customers/prospects is where every company needs to get. Your digital body language impression, coming from your heart, giving to grow, caring about adding value, making deposits before withdrawals...That's what we are talking about in this book. That is the goal of this book: To help make you aware of who you are being in the world and that in the digital world, real truly matters. Real truly makes a difference. You can be seen and interpreted, so choose to be REAL. Nothing REAL can be threatened, nothing truly unreal exists!

I am here to encourage everyone to be who they are, to be real, to give and think from your heart. Don't waste time being someone you are not. Be who you are today, right now. Don't question it. We were born with everything we need to succeed in this life. Real is best. And now that we are spending so much of our time online in a digital world, it's even more important that we are who we are and not some persona or avatar we are hiding behind in the digital world.

## MAKING IT PERSONABLE MEANS MAKING IT MEMORABLE

We all want to be recognized, acknowledged. It's part of the human condition. So how do we do that online to show we care and are listening? It's easy. Always respond to requests or reactions with acknowledgement. Thank people, write a sentence or two, show you care! This says tons about your digital presence. How you respond to a Friend request on Facebook or a request to connect on LinkedIn is part of your digital body language impression that you put out there, and its energy circulates in the blogosphere.

Let's use LinkedIn as an example. Out of the box, LinkedIn gives a canned statement when you send out a request for someone to connect with you. Rather than just send the "canned" statement, take 20 seconds and send a personal note with the request to connect. People want to know you care. People want to know you are "listening". This simple action tells about who you are as a person. Caring, too busy, thoughtful, thoughtless, distracted, self-important...You get the picture. It only takes a few moments to jot a note about why to connect or how you know each other. It shows thought and personalization. Some of us who are better at giving vs. taking personalize and care naturally. For others, it's easier to ask for and take what you want. This recovering sales shark finds is much easier to give than take. Just as the tides ebb and flow, you must be conscious of what you give, and know you have a right to receive back. It's the flow of the world!

When communicating inside of social media tools and email, it is very apparent when someone makes it personal. But sometimes being appropriately personal can be hard since business is business, and some like to draw a thick line. Social media tools are blurring the lines between business and personal. Facebook, developed for a network of college kids and their personal lives, has now morphed into a business tool. Why? Because people buy from people they relate to and like. It's about relationship marketing — "REAL-ationship Marketing©".

I challenge you to consider that your audience is on Facebook. Since the lines are blurring between social and business, it's easier to GET REAL, PEOPLE than having to think about not being who you are!!!!!

Find out a tidbit about a person AND make a more PERSONAL connection with them, like recognizing their birthday, or find out what they like, who they relate to, where they hang out. With that, you have the first step in being real and getting "related" to that person. I can't repeat it often enough: People buy from people they relate to.

## Chapter 3
## What is real (online)?

What is "real"? For me, it is just being without any thought to who I am being.

Maybe we can define what is real by answering what is not real.

I can tell you when I know I am not real:

It doesn't feel right. I am doing something, responding to something that just isn't right. I feel it in my stomach, my heart and my head. Have you ever experienced this? For me, it's not real when it's only from your head without your heart.

You might be wearing a mask for your parents, boss, teacher, sibling, partner... All to satisfy someone other than yourself.

Being REAL is really just BEING and not thinking about who you may want to satisfy with your response, what the "right" answer is for that situation. It's about having the confidence to believe in your self-worth and knowing that your voice matters, your opinion matters, you are valuable. I know from experience! So many years of not considering myself smart enough left a trail of not being REAL.

So, who are you satisfying with your career choices, your choice of partner, your choice of style, heck ANY choice? Is it a choice for the real you or someone else and their ideas and expectations?

**REAL according to Wikipedia**

**real**

1. That which can be characterized as confirmation of truth.
2. That which has physical existence.

   *No one has ever seen a **real** unicorn.*

3. *(economics)* Having been adjusted to remove the effects of inflation; contrasted with nominal.

   *My dad calculated my family's **real** consumption per month.*
   *What is the **real** GNP of this polity?*

4. *(economics)* Relating to the result of the actions of rational agents; relating to neoclassical economic models as opposed to Keynesian models
5. *(mathematics, of a number)* Being either a rational number, or the limit of a convergent infinite sequence of rational numbers: being

**REAL according to Wikipedia (continued)**

6.  (*family*) Related through a common ancestor, and not merely by marriage or adoption.

    *real father* or *real mother*

7.  (*law*) Relating to immovable tangible property.

    *real estate*
    *real property*

8.  That which is an exemplary or pungent instance of a class or type.

    *This is a **real** problem.*
    *Some say he is a **real** hero.*

9.  Genuine, not faked.

    *These are **real** tears!*

10. Genuine, not artificial.

    *This is **real** leather.*

11. (*slang*) Signifying meritorious qualities or actions especially in regards to enjoying life, prowess at sports and success wooing potential partners.

    *I'm keeping it **real**.*

**Synonyms for Real from Wikipedia**

- *(that which can be characterized as a confirmation of truth):* true
- *(that which has physical existence):* actual
- *(genuine, not faked, not artificial):* authentic, genuine, heartfelt, true

**Antonyms for Real from Wikipedia**

imaginary, unreal, fictitious, make-up, pretend, feigned, sham, staged, artificial, counterfeit, fake

**REAL according to the dictionary (dictionary.com)**

- *adjective*

1. true; not merely ostensible, nominal or apparent: *the reason for an act.*

2. existing or occurring as fact; actual rather than imaginary, ideal or fictitious: *a story taken from real*

3. being an actual thing; having objective existence; not imaginary: *The events you will see in the film are real and not just made up.*

4. being actually such; not merely so-called: *a real victory.*

5. genuine; not counterfeit, artificial or imitation; authentic: *a real antique; a real diamond; real silk.*

6. unfeigned or sincere: *a real sympathy; a real friend.*

7. *Informal.* absolute; complete; utter: *She's a real brain.*

8. *Philosophy.* a. existent or pertaining to the existent as opposed to the nonexistent b. actual as opposed to possible or potential c. independent of experience as opposed to phenomenal or apparent

## HOW TO KNOW IF YOU'RE BEING REAL

How do you know if you are being real?

What does it mean to be real?

The first check I would suggest is to check your physical reaction to what you are doing. Do you have an enhanced sense and awareness of self, i.e. heart rate, tightness, feeling "off"? These are indicators that something is not right. Your physical body is a bellwether for knowing if you are being real. Most of us don't "listen" to our physical indicators as we are buried in "noise" and "activity" pollution all around us. We also discount these indicators thinking that we must play a certain role, be a certain way, so we "suck it up" and just don't listen to our higher selves. We are so focused on our physical bodies and think that's what is "real" in us, but our higher selves are just as real, if not more so.

REAL would be in the flow, not having anything come up that doesn't agree with our digestive system, our emotional system and our heart. Practice listening to your body and its indicators.

I know I am being real when it is effortless. When I am in the flow. When "Shelley" steps out of the way and a higher source is flowing through me and just being.

Online, it really becomes a challenge to be real and listen to your bodily reaction to others. A higher tool we all have is this higher self, this higher filter. It's just a matter of tuning into it. When you are communicating online, are you "in the flow"? Are you sensitive to how your words might land on someone? Are you thinking about "am I trying too hard to sell something/someone?"

For me, many of my colleagues have told me that their perception of me is curt, short, not caring. This is really nothing new. I have been told this by friends, family and co-workers most of my adult life. I am a businessperson, and many people construe my abruptness for not caring, but this couldn't be further from the truth. However, if the goal is to be who you are without excluding anyone, then being aware of how you are being perceived is very important online.

So, we covered the physical body and its reaction to "not real". Now, you must listen to your emotional sense and see what surfaces. There is nothing that makes my emotions go off the chart like when I see an update, email or anything that is so 20th-century using cheesy keywords like "buy now", "last chance", "one day only", "act now"....and on and on. This is very old-school marketing and just doesn't work online because it's not REAL! It's not about "REAL-ationship Marketing©". So, listen to your emotions when reading about people and their business, and trust your emotional self to tell you when something just doesn't ring true.

Usually your first reaction is the REAL YOU. Trust in the REAL FACTOR!

## DIGITAL MUSCLE-BUILDER
## TO CHOOSE BEING REAL

You can tell you're starting to not be real when you are losing your self-expression, energy, vitality...kind of like when you poke a balloon with a pin and a slow leak is created. You feel deflated, down, etc.

If you feel this happening, acknowledge it, and choose again. How can you do this? Breathing helps, like the simple exercise you probably learned as a child: Count to 10, breathe and take a few seconds out to be with you and how you are feeling. Take a look at it.

Now here's the opportunity to decide what you are going to do next in the face of deflation or dilemma! Recognize the dilemma, and decide what you DO want to do.

Here is a checklist to assess if you're being real, to check your true character, to determine if you are wearing a mask and if your avatar is really who you are.

1.  Are you using words such as "buy now", "one time only", "sale", "discount"?
2.  Are you knowing in your HEART that what you are communicating isn't from your HEART?
3.  Are you asking yourself, "What's in it for them?" If you can't answer that sincerely, then you may not be coming from real.
4.  Are you giving to grow vs. giving to get? "You owe me!" just doesn't cut it out there in cyberland.


**CONSCIOUS SOCIAL MEDIA: BEING REAL ONLINE**

How do you use social media consciously to make sure you're being real online? Here are some tips:

1.  If you build an online presence, show up! What do you deduce when you go to look someone up and they haven't posted in 30 days? Your visitors will "hear" you as not participating and most likely you will not be taken seriously. Check in daily with all of your social sites.
2.  Engage with your community, and care about what they have to ask and share. Be encouraging, loving, caring, and interested in what they are up to. This is where you can truly give to grow and build lasting relationships.

3.  As I mentioned, the days of "buy this", "one day only", "special discount" are so last decade. This now becomes a major red flag when reading one's digital body language. Remember: be a magnet not a sledgehammer!

As you contemplate your "realness" online, you may have to shift your thinking. It's all about being sincere and real with your customers, clients and connections. You must care and get to know them.

> As George Bernard Shaw said, "Life is not about finding yourself, life is about creating yourself". And in this age of social networks, you can create just about anyone you want to. **Why not create you?!**

## Chapter 4
## Lessons from nature: Being real is natural

I will tell you that all of this positive energy, digital body language, coming from the heart, blah, blah, blah was not without much work and practice and perseverance. Changing behavior takes time and repetition. The key is never to give up. Never to be dissuaded by those sharks out there. Those folks who are pretending to be from their hearts, but you know they are not. They are wearing a mask. They may have thousands of followers or connections online. They may appear to be successful. However, listen to your gut. You are likely to run across some people like this. You may think you need to emulate them. You may think they are smarter or more clever than you, richer than you, more successful than you; however, LISTEN to your heart, not only your head. I have to remind myself of this less often than I used to because practice makes perfect.

AND never give up. Nature offers us some great lessons in this.

When I think about being real, I remind myself of the hard freezes in Houston in 2009 and 2010. I thought I had lost my beautiful HUGE plumeria plant. Everyone said over the course of the spring to just throw it away, it was a-goner. Yet I never gave up on that plumeria. And 6 months later, sure enough, it started growing again from the base of the plant.

The moral to the story is this: Never listen to someone and believe them without consulting and believing in yourself. You have the answers, you just need to learn to hear them, to be open to hearing them. At first, we might dismiss them, but over time, we learn to be conscious of this higher "listening" and believe in what we have to say vs. thinking others may know more than us, are smarter than us, richer than us, wiser than us, all of the "THAN US's". These all need to be dismissed.

And by the way, the moral to this story applies to all areas of our lives, not just plants! I still find myself wanting to listen to others I think are more successful than me because they must know more than I do. I have to remind myself that just because they have a lot of lemmings following them does not equal success. It's easy to fall back into not trusting your gut, your heart, your higher self. By higher self, I mean your intuitive self that guides what to do with wisdom and without over-thinking. Just be aware when you are NOT listening. Never give up on you. You have all the knowledge. Just start hearing it.

### DIGITAL MUSCLE-BUILDER
### TO HEAR YOUR HEART AND HIGHER SELF

Most of us are so busy in our minds all the time, it's almost impossible to even recognize we have a higher self, let alone hear it. And often our poor hearts sit around wishing they could share their wisdom, but the mind is being too noisy and bossy trying to run everything.

Here's a Digital-Muscle Builder to hear our heart and higher self:

Just sit quietly for a moment and remember a time you really experienced peace or joy or inspiration. Was it a special trip? A magical spot in nature? The first time you saw your new-born child? The first time you fell in love? Remember that time, and feel how your body feels. Breathe deeply and let the sense of joy, peace, happiness, inspiration spread throughout your body. Our higher self and heart like it when we are in that space. Now just be still and listen. You might get a new sense of direction from something in your environment or that small voice inside your heart that often knows what's best.

To continue with this theme that being real is actually our natural state and NOT being real is something we develop over time, I really like this blog entry by Simon Mainwaring and want to share it for its insight on how social media is connecting us to our humanity.

July 1, 2010
*Coming from the heart*

*Top ten ways social media is teaching us to be human again*
*June 27, 2010 by Simon Mainwaring*

*This week I spoke at the Cannes International Advertising Festival and one of the things I stressed in my social media seminar was the fact that "Technology is teaching us to be human again'. The two winners of the Cannes Cyber Grand Prix - the Nike/Livestrong Chalkbot campaign and the VW 'Fun Theory' fun campaign - demonstrate the truth of this.*

*What these two pieces share is the use of technology to connect emotionally with what it means to be human. For its part, Nike's Chalkbot enabled cancer sufferers and their family members to send short, heartfelt messages and have them spray painted in chalk on the road in front of the cyclists. What's more, everyone who contributed a message received back from the brand with a photograph of their message painted on the road.*

*The power of this exchange cannot be overstated. Here's why:*

1. ***IT'S PERSONAL:*** *It enabled the very personal (and therefore ultimately universal) emotions of individuals to meet the sweat equity of the athletes head on. That's a very powerful combination of heart and body in service of the cause and brand.*

2. ***IT'S GIVING:*** *The campaign benefited from free worldwide media exposure directed at the Tour de France that led to donations towards the cause and emotional comfort to other families and individuals touched by cancer.*

3. ***IT'S DIALOGUE:*** *The fact the brand replied to the contributor with a photograph of their message demonstrated that Nike was truly sensitive to their emotions and not merely leveraging them for their own benefit.*

4. ***IT'S CELEBRATORY:*** *The exchange not only demonstrated a true dialogue between brands and consumers, but it shows how a brand can successfully represent itself by amplifying the voice of its community.*

5. ***IT'S DIRECT:*** *Often brands try too hard to find ways to connect with their community. Yet this approach was so simple and recognized that what connects us is not technology but heart. Technology was put in the service of that truth.*

*The second winner was the VW 'Fun Theory' campaign in which interactive sound effects were added to physical environments to encourage healthier behaviors. This too serves as a guide to the many ways technology is allowing us to connect to our shared humanity again. In September,* ***Volkswagen*** *launched* ***www.rolighetsteorin.se,*** *a creative initiative to test if fun could change the behavior of people. The campaign has become a* ***huge success*** *in the last couple of days with a* ***tremendous amount of views*** *for the videos that Volkswagen subtly seed with this campaign. CHECK OUT THE VIDEO:* ***www.viralblog.com/online-video/ volkswagens-viral-video-serie-the-fun-theory/***

1. ***IT'S POSITIVE:*** *The ambition of this campaign was to change behavior for the better of the people involved. That's no small task yet it showed what can be achieved when you bring creativity to change.*

2. *IT'S FUN: Too often brands take themselves too seriously, ignoring the fact that people like to reveal the lighter side of themselves.*

3. *IT'S UNIVERSAL: The fact that the most popular stunt in the series has been viewed over 12 million times on YouTube demonstrates how much we have in common.*

4. *IT'S SIMPLE: Like the Nike Chalkbot work above, this campaign is strikingly simple. Yet for a brand to embrace this idea and execute it, they must recognize the importance of simplicity, human connection and technology in the service of universal values.*

5. *IT'S UNPREDICTABLE: No one could know what people would do on the musical stairs. No one could predict the exact words a daughter would write to her father suffering from cancer for the Nike Chalkbot. There's inherent risk here and powerful trust in human creativity, honesty and spontaneity.*

*The fact that these two campaigns were so highly awarded at Cannes bodes well for the traditional advertising industry and proponents of social media. It shows creatives working with technologists to foster connection through two polar opposite but equally fundamental emotions -- grief and joy. As such, these two campaigns are a master class in effective advertising and social media and Nike, VW and the Cannes International Advertising Festival deserve all the credit.*

What do you think of the two campaigns? What other campaigns have you seen that connected so emotionally?

We as human beings often struggle between everything that is inherently noble in us and our human frailties. But we always have a choice of who to be: the noble part of ourselves that creates beautiful, inspiring things or the instinctive part of ourselves left over from our animal inheritance that sinks to herd mentality. Both of these aspects are technically "real", but the difference between us as human beings and animals is that we can choose the higher path of discovering and connecting to our higher natures – a process that requires using our heads, hearts AND intuition.

## Chapter 5
## Approachability is the new handshake, trust is the new currency

So, how do we ensure that our in-person approachability translates into communicating that we are approachable online? How do we engage people so they want to hang out with us professionally when we are communicating via status updates, tweets, through our blogs and with the information we share?

Well, it's really no different than when you are with someone in person.

You want people to find you approachable. You don't want to exclude anyone. It took me a while to understand that you should be all-inclusive in your business self. Why narrow your opportunity by stating something that might exclude a segment of the population? Self-selection will occur automatically without me having to make a statement. Like begets like, so you will attract what you put out there, even online. Your approachability via the written word, your comments, your picture, what you share and what you stand for is the new handshake.

What are you portraying online? Are you coming off as judgmental, opinionated, right wing, left wing, exclusive, a chest-thumper? Ask yourself all of these questions. With more and more business brands online, it's very important to consider who you represent before establishing brand YOU.

Before you ever post about yourself using Facebook or LinkedIn or Twitter, think about how your words will "land" on the people reading them. Tell a story about yourself in the first person. Be REAL, people love to read about you in the context of real. Business is becoming more and more personal. Relationships matter. The lines between social and business are blurring with the advent of the usage of social media tools in marketing plans. Consider NOT trying to be someone you are not. NOT trying to become the wolf in sheep's clothing. NOT trying to sell because of the pressure of quotas, bills and material goods you desire. No one wants to hear about how great you are. They want you to be great, not ACT great. Make your profile information about you and who you really are. Share your likes and interests. People want to know the real you.

Let's look at some ways to think and questions to ask yourself to help you do that.

## Checklist: What Story Are You Telling Online?

*Adapted from Scott Ginsberg, author of* The Power of Approachability

You can leverage your simple idea or what you communicate online into something big and beautiful and fun and cool that:

1. Utilizes your gifts (this would be your natural-born gifts to others that come from your heart, not just your head)

2. Validates your existence

3. Gives to grow (creates abundance for all involved)

4. Makes the world a better place

### QUESTIONS TO ASK YOURSELF WHEN USING SOCIAL MEDIA TOOLS

How am I putting myself out there?

What is my unique approach?

What unique need do I fulfill uniquely?

What does my bumper sticker say?

An excellent example of approachability being the new handshake and trust being the new currency is a woman I had the pleasure of meeting. Bonnie with CEO Space lives her life from a place of giving to grow. The question she and her associates lead with is: What are you working on, and how can I help you? I mean, how simple is that? It engages the other person to share about them, and who doesn't like to talk about themselves?! It allows the recipient to let down their guard and ask for what they want. It gives Bonnie an opportunity to know someone selflessly. What a fantastic business world it would be if we all would lead with that! Come from a place of assistance and approachability vs. need and want.

I know, the bills are calling, but just give it a whirl. Take it out for a test drive. See what happens when you change your lens and see the world through giving-to-grow eyes. It's even more important online in the virtual world. We can't see you and read your body language, but your energy comes through your digital language via the written word, what you share, your picture and your approachability!

## TRUST IS THE NEW CURRENCY

Online, we certainly can't reach out and shake someone's hand and look them in the eye. So, what is out there for us to embrace? Trust. Building trust with your colleagues, customers and prospects is the new currency. The 21$^{st}$-century currency.

How do you build this online trust? Pretty much the same as we did it in the 20$^{th}$ century:

1.  Be your word. If you say you are going to do something, do it. Simple enough, but online, so many people think this online veil protects them from having to be their word.
2.  Listen to what people are saying and respond. When someone "talks" online (comments and likes and asks questions) about what you are posting, let them know you hear them. Reply back. People love to be acknowledged and recognized. This builds trust.
3.  Provide value or give to grow. Make what you give online about your Fans and followers, not about you. This builds trust.

I think a major pre-digital source of building trust was the ability to read someone's physical body language. In the pre-digital and pre-social media age, it took time to get to know and build trust with your associates. However, the same basic common sense and good citizen characteristics that in the past have always been used to gain trust are the digital trust-builders as well. It's just happening now and being built digitally. The digital footprints you leave will determine if your trust factor or digital stock is increasing or decreasing.

Social relationships on Twitter, Facebook, blogs, etc. are driving relationships that ultimately create future customers. According to my observations, these relationships generally take longer to build that same trust as in in-person relationships. (My editors have shared that they've had the opposite experience, so as with everything else on social media, people's interpretations and realities will be different.)

What most people can agree on these days is that the old sales approaches don't work anymore. Scripts, canned speeches, talking features and benefits and give to get really don't bring the results we might hope for in the new digital age of "REAL-ationship Marketing©". We need to be digitally available, engaged with our colleagues, prospects and customers and approachable with a new listening for what they are sharing and asking.

## Chapter 6
## Getting real: Dropping the hidden agendas

This book is designed to open your **awareness and recognition** of people's online footprint, their digital body language. This is most important for adults and a screaming desire of mine for our youth. So much is out there and believed and yet a lot of it is fabricated, manufactured for the pure purpose of selling. One of my goals is to bring awareness to what is real vs. salesmanship.

Being SMART, TRANSPARENT and CONFIDENT online is easy unless you are being motivated by greed or popularity or what others tell you is right.

Reading someone's digital body language is not an easy thing. There are many wolves in sheep's clothing; that's what got me started on this path in the first place. What's important is your digital body language, that you're being real and that you're coming from integrity, honesty, transparency and confidence. How hard is that?

It's not! It's as simple as being honest, transparent and coming from a place of integrity.

But if you have to think how to be those things, then you are not those things yet. The good news is that they can be developed.

**EXAMPLES OF HOW TO BE SMART, TRANSPARENT AND CONFIDENT**

1. Avoid coming from fear and a scarcity mentality.
2. Avoid caving in to pressure and other people's expectations.
3. Know that you were born with everything you need to do what you want and be who you want to be.
4. Avoid letting people you think are smarter, richer, prettier or more experienced influence your decisions; know that one of your best potential barometers is your heart.

Getting real when getting online has to start with yourself. Since like attracts like, you really have to know who you are and what you want to represent when setting out to build a community. People will only trust you when they sense or know you care about and serve others. But the irony in this is that the only way you can know how to do this is to be good at caring for and serving yourself (not self-serving, but loving-serving).

So many of us have the negative tape recorders running in our heads that block our ability to really connect with others. So many of us have built up inhibitions over the years that little Bunny Shelley didn't have when she was hip-hopping her little heart out on the stage until she got zapped instantly into the adult world of inhibition.

So let's start with ourselves. Let's learn how to become our own LIKE button! And then we can SHARE (as with the Facebook SHARE button) with all we encounter.

## DIGITAL MUSCLE-BUILDER
## TO BECOME YOUR OWN "LIKE" BUTTON

Caring people – that would be you since you're reading this book! – usually have a harder time nurturing themselves than others because they're too busy putting others' needs first. Just think of teachers or nurses.

To be real, we have to like ourselves – to become our own "LIKE" button. No matter how much we might like others, if we don't like and love this self we're walking around in, we won't be able to be completely real.

So here's the Digital Muscle-Builder to become our own LIKE button:

Write down all the things you like about yourself, yes, ALL OF THEM!  You're not allowed to even THINK for a moment about anything you don't like about yourself; chances are, you already do that far too much in comparison to the time you invest in reviewing what you like and what others find likeable about you. Do this at least once a month, and especially when you are feeling down on yourself for some reason or when you feel like you aren't being real.

When you have the attributes of integrity, honesty, transparency and confidence, you will attract like-minded prospects, customers, colleagues and friends. Like attracts like! Any HIDDEN AGENDAs naturally drop away along your path.

### Physical Reactions to "Not Real" or Hidden Agendas (in yourself or others)

- Heart speeds up
- Eye contact is not maintained
- Breathing becomes shallow
- Pulse increases
- Heavy pressured or constricted feeling in chest
- Nausea or vomiting
- Neck gets stiff
- Backache
- Irritability
- Sweaty palms
- Impulse to flee
- Intestinal gas and/or diarrhea
- Heartburn or acid reflux
- Head pressure or headache
- Sense of unease or malaise
- Butterflies in the stomach

*source: energy arts alliance*

Once you're attuned to hidden agendas in yourself or others and are more in the space of true realness, you start asking yourself, "How do I use social media responsibly?" "How do I ask my workshop attendees for referrals responsibly?" If you give enough, you deserve to ask in return. I have a hard time asking and receiving. I can give all day long, but when it's time to ASK for help and RECEIVE the help, well, that is work for me. I do know I deserve it and that's what I will attract through my digital body language.

The reality is the clearer and purer your intentions and heart, the more you cannot be deceived. Deception is impossible when your intentions and heart are clear and pure.

When your heart is fully engaged in who you are, what you do, how you are being, then you will be perceived as coming from the heart. The heart is a HUGE filter and regulator... We must learn how to listen to it. That's the part we all need to practice. Listen to your heart; it's a communicator with your higher self. And it's almost always right.

## DIGITAL MUSCLE-BUILDER
## TO COME FROM THE HEART

Sometimes the stress of everyday life can disconnect us from our hearts. Or maybe we didn't grow up in a very emotionally warm or supportive environment. Or maybe our hearts are so sensitive we have to put them behind a wall to avoid getting overwhelmed by feeling or absorbing too much from others or the environment. Whatever the reason, if we aren't in touch with our hearts, we'll often appear insincere or unreal to others, and it will be difficult to experience a sense of security, peace or contentment. Security, confidence, peace and contentment are magnetic qualities, qualities that make people want to be around you. This is true for personal life and true in business.

So here's a practice to come from the heart:

*Close your eyes. Remember when you did something kind for someone or when someone did something kind for you. Now be aware of the center of your chest. How does it feel? Does it feel warm? Does it feel like it's getting full or like it's overflowing? Do you feel a pleasant flutter?*

Most of us will feel at least one of these sensations. If you don't, it might be that your heart is out of practice or atrophied, just like what happens to our muscles when we don't exercise regularly. Just keep practicing, and one day, your beautiful heart will wake up and start speaking to you. You're human, and that's what every single one of our hearts is meant to do!

Everything about your online behavior sends out a message to other people. It's like throwing a stone into a pond. That first ripple creates more ripples into infinity. It's the same online. Companies cannot afford to bury their heads in the sand with social media. All you need is one stone thrown to ruin your business, your reputation if you don't respond or don't respond properly.

The good news is you can clean these kinds of situations up just easily. (Remember the earlier story of the pizza chain that used Twitter to reverse the effects of a potentially harmful viral video?) Create extreme customer service and raving fans by addressing their comments and monitoring and responding to what people are saying about your brands. Do you like what they are saying about your brand and "brand you"?

What's your reason for being out there using social media tools? That is what people will get about you. Do you want to: Generate leads? Generate community? Generate sales? Generate customer experience? What are you leading with? For example, Facebook is world visionary, with the stated mission of giving people the power to share and make the world more open and connected.

**DON'T HIDE IN THE WEEDS; BE AUTHENTIC!**

Content strategy is important. Stay engaged with your customers. If someone is so into their own brand or company, they may think people want to hear random, not terribly engaging things about a brand or company, but that is not the truth. Communicate the most interesting things YOUR COMMUNITY wants to hear about. They want value for the content they are consuming. You have to step back and be realistic about where it fits in with your customers' lives.

Add value out there. A good rule of thumb is to add 80% value with helpful information and no more than 20% promotion about you or your product. What you share can be in the form of links to blogs or articles of value, video clips that are educational or just make you laugh or references to other people that add value with the goal of becoming a human aggregator for your field.

Experiment and find your style, what people relate to. What's your style? It just needs to be true to you and your heart. Review what qualities you want to make sure you experience and bring to other people every day, and those qualities will pretty much point you to your style.

You can evaluate your analytics and look at the numbers all day long, but the bottom line is ROI. Not the ROI you might think, but rather, **"Return on Influence"**. Giving to grow is the ultimate investment, and your ROI will ultimately give you a return on investment.

As we evolve as a species, our concern will grow into more about our return on influence vs. return on investment. If we truly are being in our right mind and selves, then we will not concern ourselves with the return on investment version of ROI. The traditional

Return on Investment holds the old "give to get" approach in high esteem. But how do we measure our Return on Influence? How do you value the ROI of your cell phone? Pretty tough to do. It is almost a matter of trust in knowing you are coming from this place of giving and setting an example of being in your right mind, knowing that this way of life will ultimately reflect your return on influence. This is part of evolving from a "me"-centric way of being to realizing that you are building a digital community that wants to receive value from you. Once the value is delivered, the Return on Influence is achieved. Thus, the more you give, the more you grow, the more influence you have on your community.

Be your word. Don't over-commit and then fall short. Under-commit and over-deliver.

Over-committing and under-delivering is a bad habit often developed in the business world. For example, when I was in sales, every quarter we had sales quotas to achieve. When it came down to the last week to make quota, I observed other salespeople over-committing just to get the deal. They said the products did things they couldn't do. I could just never do that.

## PUTTING YOURSELF OUT THERE – RELATIONSHIPS MATTER

Your LinkedIn profile and Facebook Business Page need to represent you as a person or people, not strictly as a business. Make them personal. People buy from people, not from companies. Speak in the first person, make content relevant and in line with your values and company values. Experiment with how often you want to post. Keep a backlog of creative posts. You never want to post or respond when you're not feeling it. People can read this energy, so don't post anything that doesn't come from a positive, energetic place.

We walk a fine line with how often to post information. Posting five times a day may be a bit much, unless you are on Twitter and communicating short bursts of valuable information over the course of the day (what your followers deem as valuable, not you!). You risk over-playing your hand. Always ask yourself, What's In It For Them, and is it relevant? Always mix up the type of content you put out there. DON'T have HIDDEN AGENDAS. Remember, most people can see that wolf in sheep's clothing, so be transparent.

One way to determine what is of value to your customers is to simply ask them. Create a poll for them to answer questions that will assist you in creating the value they seek. An example of one of the polls I did on Facebook was about the time of day my Fans were on Facebook reading the newsfeed: 1) early morning 2) midday 3) evening.

Another means is to view the response you get when sharing information on Facebook. How many people selected LIKE to something you posted? How many comments did you get by your community on a Facebook posting?

On LinkedIn, you can use the same method to determine the value. If you are inside of a group and share something, did others comment and become engaged with you? Did they receive value from you? So often we forget to just ask our community what they want and what they value.

If you are out to make your customers happy, don't be self-focused. Have a sense of awareness and recognition with a sense of instinct and impression. Seek interaction, be a good customer champion and a great customer service person. That helps your authenticity come through...

**READING DIGITAL BODY LANGUAGE ONLINE**

As we said earlier, reading someone's digital body language is not an easy thing. When we're in person, we can tell if people are engaged by looking us in the eye with interest. Or leaning in towards us. Or smiling with their mouth or eyes. We can tell they're not connecting if they keep looking away. Or cross their arms. Or interrupt us as we're speaking or – worse! – text while we're speaking!

Online it's a bit more challenging since we cannot read their body language, i.e. facial expressions, bodily reactions, etc. We don't even have the luxury of hearing their tone in their voice (although after interviewing many people, there is a bit of a "heads up" on interpreting the written word to read emotion. See interview Chapter 7 for more on this.)

Here is a map we developed to compare in-person body language cues and possible digital equivalents. This set of digital equivalents is a compilation of some of the best and clearest answers we received from interviewees.

| In-Person Body Language | Digital Equivalent |
|---|---|
| Maintains eye contact | Responds in a timely manner |
| Listens vs. talks | Gives positive, interested feedback |
| Language cues, asks questions | Same online; engages you |
| Arms crossed - frustration | Using capital letters to stress getting a point across |
| Texting while talking to you | Ignoring you when asking questions; not responding to posts or comments |
| Someone mirrors or mimics you | Similar online; you see people using your words or style |
| Nervous or skeptical gestures: playing with hair, playing with glasses, squinting | Delays in stream of conversation during real-time communication (i.e., chat, instant messaging) |
| Fabricates answers to questions | Buries point under mountains of unnecessary words |
| Tone of voice | Use of punctuation and emoticons |
| Confident | Straightforward, friendly |

| | |
|---|---|
| | language without a lot of boasting or credentialing |
| Approachable | Invites feedback and asks instead of making statements or declarations |
| Handshake | Inclusive (invites others to share and contribute) |
| Loud, boisterous verbally and non-verbally, pushy | Over-communicating, too many emails, interrupting or changing flow of the conversation, sharing too much information |
| Negative and diminishing words and phrases (directed to oneself and/or others) | Same online |

*Chart compiled by: energy arts alliance from people interviewed by the author*


## MAKING IT REAL BY LISTENING TO OUR HEARTS

WHEN THE HEART IS FULLY ENGAGED, the heart is a filter. When it feels wrong, it is wrong. I can't tell you how many times I've experienced doing something and immediately having a physical reaction to the deception I may be imparting, the white lie I am telling, the fabrication I am stating! I know when something is "wrong" here, meaning, it truly is NOT who I am by nature. We are so pressurized to achieve and be more that we often are not "listening" to our bodies tell us what is right vs. wrong.

There were so many times in the past that I used to say what I thought others wanted to hear. It was the safe place to be. Now, as I become wiser, I realize via all the alarms in my physical being when something is not what I should be doing. The saying, "the TRUTH will set you free", is liberating. You know in your heart when something is right. Now, if we could just listen to our hearts in our daily lives and trust in what we are feeling. To get out of our heads just for a few minutes every day is the start of a path of liberation and freedom into who you were born to be...honest, real, caring and free.

In so many sales meetings, I was so judgmental of myself that I would rehearse in my mind over and over what I was going to say before I opened my mouth to say it. Anytime I rehearsed, I wasn't saying who I really was. Another example was when I would have to do a presentation. I had to memorize what I was going to say instead of having the faith and confidence to just be. The ultimate lesson that I finally got was, know your material, but just get out of the way and let your higher self communicate. That's what people fall in love with. That's what people listen to. That's what people want more of. That's what people relate to.

## WHAT ARE YOU WEARING ONLINE?

We spend a lot of time wondering what appropriate attire is for various occasions we go to. Whether it's a wedding, business meeting, casual Friday, we always wonder what to wear. Do you ever think about what to "wear" online? Whether you are communicating with friends, business associates or family, your digital footprint follows you wherever you go.

We give a lot of thought to our hair, makeup and wardrobe when we venture out into the world. Billion-dollar businesses exist because of it. Most of us would never consider showing up to a breakfast meeting or networking event in our shorts, T-shirt and baseball cap (although I have been guilty of that once I know who I am meeting with and they know my heart and being).

So why would you consider "going out" online and not having your best intention on, your higher self shining through, your honesty coming first?

**DIGITAL MUSCLE-BUILDER: PLUGGING IN TO GET YOUR HIGHER SELF SHINING THROUGH**

If you really want to get connected with your higher self, which is the best friend all of us are born with, a very helpful tool is the online "Plugging In" Audio App from my energy advisors at Energy Arts Alliance. To get plugged in, go to **www.energyartsalliance.com**.

We all hit that enter button way too fast without even thinking about the impression we are making. Through gathering information for this book, I was made aware that I come across as abrupt and non-caring. That was a bit shocking to me, although, I do know that is how most people interpret me in person. Why? Because I am direct and don't spend a lot of time on the "fluff". The "how are you? How was your day?" I basically go right to the heart of the matter and deal from there. Now many people appreciate this, but not all.

You need to be aware of who you might be alienating when you are online and communicating. If your goal is truly to exclude no one, then you need to decide on what you are going to "wear" online and how you are going to be perceived.

**DIGITAL MUSCLE-BUILDER**
**TO KNOW WHAT YOU ARE WEARING**
**ONLINE**

To discover "what you are wearing online" or how people are perceiving you online, here's a survey you can post on your Facebook page using a third party polling app or the Question feature:

How do you perceive me on Facebook?

   a.   Caring
   b.   Not Caring
   c.   Stressed Out
   d.   Listening
   e.   Too Busy

Get creative, have fun and come up with your own descriptives. You may be surprised by what you learn.

No matter where we go, we are so dressed in "looking good", and online it's even more important to "come from good" since it's difficult to look good on all occasions.

The bottom line is this: When you are online, try COMING FROM GOOD vs. LOOKING GOOD. As we stated earlier, your energy will be read online even if we can't see what you are "wearing" or how your hair is done. In fact, like some people whose other senses heighten when they lose sight or hearing, we may gain through our evolution as a "digital" species a better sense of what people are "wearing" – from the inside out – by the energy that they project via social media.

If you can truly come from your heart, that amazing filter we have for right and wrong, then you will always be in your Sunday finest and the impression you make will be well received and reciprocated!

Nonetheless, be aware of your "look" on the various social media tools that you use to communicate. This IS your digital footprint, your digital impression, your digital body language. Be aware!

**ARE YOU A SALESPERSON OR A RESOURCE?:**
**THE POWER OF GIVING TO GROW**

Remember, it's about giving to grow and truly giving to grow. If we all could forget the GET part, how our lives would become richer as well as the impact you would make on others' lives!

Let me share some examples of this philosophy, how I apply it with social media and how it works for me:

First, I truly do answer questions and create videos for anyone who needs help using social media tools. When I am doing this, I am truly coming from a place of GIVING. How does this help me grow? Well, so many people share my training videos and links, and in turn, a lot of what I offer as my GIVE goes out into the world and GROWs via new Fans, prospects and clients. On LinkedIn, I have found that the way to receive recommendations is not to ask for them, but to give them.

Since I have become aware of this philosophy, I am much less anxious about what's in it for me vs. what's in it for them. I still have that old sales shark rear its ugly head every now and then, however, practicing and exercising the "give to grow" philosophy has come back to me ten-fold in new clients, fans, prospects and more importantly, a fantastic way of conducting my business.

In thinking about applying this principal, I try every day to provide some social media resource for others to learn from. So whether I am sharing a link to a great blog or article, introducing a colleague to someone that they need to meet, creating a video to answer a social media question or sharing a joke to make someone laugh, it's about engaging with people and enriching their lives. Make it easy for people to connect to you both personally and professionally. Engagement. That's the answer.

What does your digital body language say about you? Are you a salesperson or a resource for your friends, followers, connections? Are you always reminding them what you do? (Like they don't already know?) Now don't get me wrong: It's important to stay in the minds of your associates, clients and prospects. Just shift your perception a bit and see yourself as a resource and engage through giving to grow. They will remember what you do and what services you provide.

Secondly, do you have CUSTOMERS or customers who are true Fans? This is a terrific reference point to measure the impression your digital body language is leaving. Remember, a major goal in life is to make a difference, come from our hearts, give to grow...If you live by those principles and you are REAL, genuine, transparent, I guarantee you that you will have Fans. Not just customers but raving, Super-Fans who speak positively about you and the service, product that you provide. AND those Fans will tells others about "brand you" and on and on. You will go viral!

This is the power of being genuine online. If one of your Fans has 130 connections, and they tell those 130, and then those 130 tell their connections, you can easily see how your name/brand/reputation will go "viral". This is the power of social media and the ability to grow your brand through giving to grow. Fans go beyond being satisfied with your product/service. They have got to have you. They want to drink from your fountain. The more real Fans you have, the less selling you will ever have to do. You become a trusted advisor and resource, not a salesperson.

My experience has taught me over the years that just BEING from my heart and giving to grow, people want more of that. NOT ME, THAT. What's THAT? "That" is the light that shines through me and shines in all of us. It doesn't come from my head, it's from my heart and my heart is pure light, pure energy, pure. We all have this light, this higher purpose, spirit, call it what you will. People are drawn to this like moths to a flame. It's their essence and no matter what you are selling, if you are aware of being real, you will always benefit and grow.

*Not so likely to attract people to what you have to offer!*

Ok, no getting too woo-woo here. The point is, I have always attracted people to me. It's not ME that attracts them, it's my pureness of heart. And we all have this. We just have to learn to trust it. And we have to practice, practice, practice until it becomes automatic.

As I discussed in a prior chapter, I discounted this heart stuff while I was a sales shark, didn't listen, until the day I walked out of that Fortune 500 company with the pure simple mission of bringing spirituality to corporate America….Yup, now that was one big mission. The two go together like oil and water in most large companies.

Yet after spending a couple of years writing and delivering a course I developed called "Living Life Fearlessly", I realized I don't need to bring anything to anyone. This was a huge lesson for me (and it was very liberating I might add). I need do nothing. What I realized is I just need to BE. I just need to TRUST in my authentic self, whether in person or online. Be the spirit I was born with and shine that light wherever and whomever I am with.

How can we apply this in social media? In every status update, post on your Business Page, comment on a blog, consider who you are being. Are you coming from this higher energy, or do you have a hidden agenda? DON'T PUSH THAT SEND BUTTON until you ask yourself this question.

Let's explore some phrases I've used and others I learned from Scott Ginsberg, Mari Smith and CEO Space that you can use to bring your realness when you interact in-person and online. These will leave a positive, giving, digital impression with those who are interacting with you and those who are lurking (meaning, they read what you say and put into action what you recommend, but you may never know this is occurring because they don't respond or let you know they are there).

An example of lurking: I posted an instructional video on my YouTube channel. Hundreds of people viewed that video, yet it only received a handful of comments and ratings. So, the lurkers are out there, and there are a lot more people who see you and the digital impression you leave than you may even know about. Just another reason it's even more important to GET REAL, PEOPLE!

## PHRASES AND APPROACHES TO HELP YOU BRING REALNESS ONLINE

*Is there something else I can help you learn?*
*(Scott Ginsberg, http://www.hellomynameisscott.com)*

This is unexpected and thought-provoking and revolves around you wanting to educate clients, customers and prospective customers. Are you handling your customers or teaching them? Comcast took a bunch of lemons on Twitter and made them into the sweetest of lemonade. A lot of companies are realizing the ability to use social media as a customer service tool and take complaints and issues and leverage extreme customer service to make wrongs into rights. Whether it is Domino's Pizza creating a video to apologize for a mishap or Comcast responding to issues promptly and willingly, social media is a fantastic means of creating a great customer experience.

> *"I have only one mouth, but two ears, and I'm listening*
> *to you intently."*
> *(Adapted from Mari Smith, www.facebook.com/marismith)*

The goal of this statement is to lay an immediate foundation for fantastic personal service! What does it say about YOU and the impression you leave on people when you make it all about THEM?

Mari Smith, source of the quote above, is an example of someone who has taught me extreme customer service (and a whole lot about Facebook). She is a great example of attentive service with thousands of Fans on her Facebook page. She exudes openness, personability and fantastic customer care and service. Her digital impression is that she is caring, concerned and makes it "all about me". How does she do this? She simply "listens" on her Fan Page and responds in a timely fashion. She also is very encouraging and engaging, and her picture is open and inviting.

I have never met Mari, however, we have communicated via social media for a while now, and when I meet her in person, it will be as though we have known each other for years. Engagement... that's the key to succeeding with social media – "REAL-ationships" = the new marketing.

*What are you working on, and how can I help you?*
*(Bonnie Karpay, CEO Space, www.ceospace.net)*

I have mentioned this elsewhere in the book. This is what CEO Space uses in all their interactions. Again, give to grow... Nuff said!

*I am sure we can find a solution.*
*(Shelley Roth, **www.facebook.com/SpringboardWorks**)*

This immediately puts the fan, customer or client at ease. It's coming from giving to grow and is positive and flexible. An example of this: I had a Fan on Facebook who was very stressed out because she couldn't figure out how to delete status updates that she didn't want on there. Responding to her in a timely fashion and assuring her I would help her do this put the Fan at ease. I can tell you that this Fan is now a "Super-Fan", and when someone needs help on Facebook, she refers them to me.

**Chapter 7**
**Reading and perceiving digital body language**

Digital body language...What is it? How do we interpret our fellow travelers online? If we can't see it, how do we know what we're reading? How often do we read our own interpretation into someone or something that is misconstrued and not intended to be heard by our own "listening"?

So many questions, so much time being spent on communicating digitally with friends, colleagues, business associates, family, bosses, employees... It can be a bit overwhelming. If you're anything like me, or how I used to be when online, I just took everything at its surface value and "believed" what I read, listened to, saw. Now, with the world going digital, I am much more concerned about the state of the union and what we truly are getting and giving online.

This led me to interview various people from different walks of life – an artist, a religious scholar, salespeople, a corporate executive, a scientist - and share with you the results of my mapping exercise. We took everyday physical body language and mapped its digital equivalent based on the feedback we received.

This process gave us a basis for verifying that in fact we are putting out energy online through our digital impressions, which are interpreted in many different ways by right-brained and left-brained people. The bottom line from this research was to help others be aware that REAL is REAL whether online or in person and to be very aware that the digital impression you're leaving on others is how they "see" you.

Let's start with the obvious, the picture that you post online. It is the closest we get to "seeing" the real you. Among all those we interviewed, it was agreed that a picture puts out an energy and thus an interpretation of who you are. Is your picture inviting, asking people to come learn more about you and what you do? Are you smiling? What makes you want to learn more about this person from their picture?

Smile. That's the number one input received. A smile is inviting and welcomes people in. It gives us a comfort level and feeling of engagement right from the start. Before we read anything about a person, that smile is welcoming. So, SMILE! A picture DOES tell a thousand words!

What I learned from this survey exercise is that everyone has that sixth sense about what is real online for them. They may use different indicators; however ultimately, they do experience alarms going off when something or someone isn't authentic. (Being aware of and listening to these alarms is the key!)

Take a moment and reflect on what some of the things are that have happened to you online that weren't **REAL,** from the heart, transparent? When I am online, my immediate reaction is to the person's picture and the sense I can get of who they are and what they are up to. Then, going beyond their picture, if what they are saying is all about them... red flag! Or if they are trying to sell something... red flag! Or if they appear to be overly anxious to please... red flag! I can go to someone's Facebook profile/Business Page and LinkedIn profile and get a good feel about the type of person they are. The bottom line is that if it's all about them... red flag! If it's too good to be true, it most likely is.

I will tell you that every person interviewed agreed that you will attract what is real by what you project. You need to be wise and be your real self, BUT ONLY share your real self in a way that is useful and not alienating for others.

**SOCIAL MEDIA TOOLS ARE NEUTRAL**

All social media tools are neutral. They have no right or wrong, no good or bad, no high and low in and of themselves. It's we as humans who make them used for good or bad, right or wrong, positive or negative. In my opinion, social media tools are to be used for EVOLUTION, not INVOLUTION...for good, not for bad. To help us grow vs. regress. If Mark Zuckerberg, founder of Facebook, lives his mission, Facebook will be used to give people the power to share and make the world more open and connected.

How we end up communicating and what influence we bring as humans to this amazing tool, only we can decide. If we consider using these tools only for a higher purpose, for giving for growth, that is the next step on our paths.

Let's face it: It's very easy NOT to be real while online. You can wear so many masks from super salesperson to beautiful diva to successful financial wizard. Creating a mask on social media is easy. The hard part is deciding to take the mask off.

Is it hard for you to just be you? No layers, no agendas, no outcomes. Do you wear a mask? We all are sisters, brothers, daughters, mothers, fathers, employees, friends, coworkers. Wouldn't it be great to just BE who we are in all situations without regard to what mask you will wear today? Wouldn't it be great to approach all the roles we have in life with the basic core of being REAL? Isn't who we are one and the same?

You have an opportunity to translate your REAL self in the physical world, not just in the digital world. You have an opportunity to be real in your digital body language, and you also need to be aware of who is being real/transparent and who is minimizing themselves or being a chest-thumper.

Social media tools give us an opportunity to learn about ourselves in all areas of our lives: Are you concerned with the NUMBER of friends you have vs. the QUALITY of friends you have? The NUMBER of business associates vs. the strategic business associates? What does that tell you? And what does it do to people who already compare themselves too much to others and beat themselves up for not having enough Friends on Facebook compared to their peers...? (By the way, the average person has 130 Friends on Facebook and 125 connections on LinkedIn.) And let's not even get started on social media bullying. That's the topic of a whole other book!

The long and the short is that social media tools are neutral reflectors of who we are. They can help us or diminish us, depending on how we interact with them. They're also viewed very differently by different natures and temperaments of people.

It's only logical that people who are already naturally social and extroverted in the physical world will be star social media users. More introverted people may simply avoid social media more, but that would be a mistake. Without applying the self-pressure of "keeping up with the Joneses", social media outlets can be amazing opportunities for wallflowers to come out of their shells and shine their light in the world.

So how do different people in different professions view social media and digital body language? We interviewed 25 people to find out. We have included the interview tool and digital body language map tool we used here. See Appendix II for a sampling of complete Digital Body Language Survey Tool and Digital Body Language Map Tool responses from these people representing a range of different professions and industries.

The survey asked respondents to answer questions about how they assess and interpret realness online. The Digital Body Language Map gave current in-person body language cues in the left column and asked respondents to fill in their version of digital equivalents in social media in the right column.

## Digital Body Language Survey Tool

Here's the actual survey we asked people to respond to:

*We are bringing a new perspective to interpreting "digital body language" and "digital footprint" for online communications, including social media applications. Be part of a new frontier and give us your thoughts.*

*Your input is valuable and may be used as part of a survey included in the* **GET REAL, PEOPLE!** *book and* REAL-O-METER<sup>©</sup> *application currently under development. Please answer as honestly as you can.*

| Digital body language Survey Tool |
|---|
| What, in your opinion, does it mean to be REAL? |
| When you meet someone, how do you know if they are being real? What are some examples? What are some physical cues? Emotional cues? Mental cues? Language cues? How about online via blogs, Facebook, company website? |
| What concerns did you have prior to using social media tools like Facebook and LinkedIn? (check all that apply):<br>\_\_\_Privacy<br>\_\_\_Security<br>\_\_\_Spam<br>\_\_\_Not sure how to brand yourself/your company<br>\_\_\_Will people understand our brand or misinterpret it?<br>\_\_\_Other (please elaborate) |
| Would it be helpful if you could "read" people's digital body language? If yes, can you give examples of why? |

| |
|---|
| What tells you when someone is not being: REAL/SINCERE/HONEST/FORTHRIGHT/TRANSPARENT online or on social media platforms? What are some indicators of "deception" or deceitfulness?<br>___Used car salesman syndrome-chest-thumper<br>___Personal profile picture-physical impression<br>___Energy from profile picture-intuitive feeling<br>___ Photo albums or videos<br>___Long, goes on forever, "sales pitch" via email etc.<br>___Buried pricing, have to search high and low to find price<br>___Buried call to action<br>___Shameless promotion; chest-thumping<br>___It's all about them<br>___Someone that asks me nothing about myself<br>___Agenda is apparent<br>___Hidden agenda-smooth marketer<br>___Ask for what they want vs. what's in it for me; what's in it for them vs. what's in it for me<br>___ Other. Please elaborate |
| How do you read someone's digital body language? |
| How do you read a company's or organization's digital body language? |
| How do you know if they are being "real" online (according to your definition of real)? |
| What are some indicators of "realness"? |
| What are some indicators of "deception"? |
| What alarms are set off when you view:<br>• A LinkedIn profile page?<br>• A Facebook Fan Page?<br>• Twitter home page<br>• Blog pop-ups<br>• Website pop-ups |

> What is something that happened to you online that just didn't feel REAL?

## Digital Body Language Map Tool

Here's the digital body language map that we asked people to fill out to help us identify what people perceive as digital equivalents to in-person, physical body language cues.

| In-Person Body Language | Digital Equivalent |
|---|---|
| Maintains eye contact | |
| Listens vs. talks | |
| Language cues, asks questions | |
| Arms crossed: frustration | |
| Texting while talking to you | |
| Someone mirrors or mimics you | |
| Nervous or skeptical gestures: playing with hair, playing with glasses, squinting | |
| Fabricates answers to questions | |
| Tone of voice | |
| Confident | |
| Approachable | |
| Handshake | |
| Loud, boisterous verbally and non-verbally, pushy | |

All of the completed surveys and mapping tools we received appear in their entirety in Appendix II at the back of the book. Here's a sampling of responses we found most informative, interesting or original and trends we identified among almost all the interviewees.

- "In various countries colors represent different things. In America, white represents clean, purity whereas in India, white represents death. From a marketing perspective, the only universal color for every country is blue. Depending on the shade, it represents anything from royalty to calmness, tranquility, peacefulness. Inappropriate colors can have a negative effect (e.g., a yoga place that's bright purple and black)." – Marisol Graham, Principal, Baxter Graham Design
- Curse words in social media turn people off.
- Interviewees showed a perception that spelling and grammatical errors can be an indication of not caring. This might not be fair or the truth (especially for non-native language speakers), but it's a perception that's important to be aware of.
- If someone responds to you, it's like the digital equivalent of maintaining eye contact.
- People who are negative, whining and complaining online are no more popular in the digital world than they are when they behave that way in person.
- Definitions of being real:
  - Our scientist, Raoul, defined real as: "Transparent to a level that is no fears,

no self-serving motives. My visual picture of transparent being a little bit of a Star Trek fan is a Vulcan mind mild and truly getting into each other's brains; that is transparent."

➤ Designer and entrepreneur, Marisol Graham: "To be truthful and honest about who you are, where you come from, what you want to attain…"

➤ IT Consultant, Diversity Trainer and Team-Builder, Becky Earl: "To be clear, concise in one's communications and one's presence… No personas if we can help it."

➤ Religious scholar, Dr. Jill Carroll: "To know the truth about yourself, to know what's true about you, for you and to share that without tricks, strategies, etc. To know the truth and speak it."

➤ Our anonymous artist: "Doing things without influence of others, societal influences or peer pressure – more easily said than done."

➤ Strategic Business Consultant, Ruby Renshaw: "To share from a place of self-acceptance such that there is no fear of someone judging who you are. To tell it like it is right from the beginning without a hidden agenda."

➤ Business Owner, Kellie Schneider: "To be authentic, to follow through and do what you say you're going to do."

➤ Lilly: "A real person does not intentionally or consciously represent themselves to be anything that they are not."

> ➢ Business Owner and Networking Queen, Marian LaSalle: "I find that the phrase Energy Arts Alliance came up with, "give to grow", is exactly what I look for in myself and the people I meet. I want a giver's gain attitude in the people I do business with and when choosing my close friends. To be <u>real</u> is to care about more than you. Under normal circumstances, you can tell fairly fast if the person you are meeting is real."

It was surprising to me when I did the survey that people really did NOT have trouble discussing their experience with digital body language online. Of course, the obvious comfort zone was the photo and all CAPS letters as leading indicators of digital body language. But they didn't have any trouble going deeper into other digital body language cues. People are very interested in how to do better reading people's online behavior, so it would make a lot of sense for a scientist or sociologist to do a study on how to measure digital body language quantitatively.

My conclusion after this exercise was that you're never going to know how others may be interpreting you unless you ask yourself these questions first:

- How am I being interpreted by others?
- How are my picture and words landing on people?
- How are my grammar and punctuation?

It was obvious from this interviewing and survey exercise that people's body language can be interpreted online. Whether through a smile, the written word or pictures that are posted, we all formulate opinions of each other from digital input.

The key to interpretation is to be aware of what your senses tell you when dealing in the digital world, and always consider how you will be interpreted online.

The bottom line is to be conscious of your digital body language. It doesn't hurt to ask people what their impression is of you. Don't be afraid to ask. If you don't want to know, you don't want to grow!

## Chapter 8
## The nuts and bolts of getting real online: Specific ways of using social media effectively

Now that we have discussed the importance of being who you really are, let's take a look at the most popular social and professional networks online with practical ways and strategies of using them effectively to express your real self. We'll look at the top social media platforms of Facebook, LinkedIn and Twitter. As we discuss using these tools for brand promotion, customer service, growing your prospect base, meeting new colleagues and establishing your identity, keep in mind that even though we have the temptation to portray ourselves and our businesses in a way that might not be authentic, real ALWAYS wins.

It really is about being conscious and socially responsible. It's a trust economy and people buy from people they trust, have something in common with and whose digital body language radiates truth. Trust is the new currency.

At the heart of social networks are communities. In all communities there are people with high integrity, questionable integrity, honesty, dishonesty, givers, takers, wolves in sheep's clothing, self-promoters, sales sharks, manipulators, etc.

Let's take a look at how your digital body language is represented in the social network communities of Facebook, LinkedIn and Twitter.

In each of these social network communities you will find platforms for joining groups, exchanging ideas, offering opinions, participating in discussions, answering questions and other means of sharing information about who you are and what you represent.

People collaborate and share information, ideas, and solve problems as community members.   People become a part of the community by participating. This is important because without participation you can be viewed as a "lurker". A lurker means you watch and listen, but don't contribute.  People in these networks would then not consider you an active part of the community if all you are doing is hanging out and lurking. It's all about relationships, "REAL-ationships".

So, out of the gate, when you join groups, participate in discussions and consider this: You want to be with others with similar interests and share ideas and feel part of the group. You can come and go at will, but building relationships is key to establishing trust and goodwill. Members are more likely to share their feelings and express their true opinions once comfortable within the group. Build relationships before you need them.

The key to all of the communities online is **ENGAGEMENT**. You want to create conversation where everyone is participating. You want to encourage feedback and sharing and build your communities organically. The content is provided by you as well as any other source that adds value, including all who are engaged within that community.

The goal of any social network is to take it to the next level by creating a community where people are sharing, giving, growing and engaging. That's how you grow. That's how you begin living the mantra, "give to grow".

## GUIDELINES FOR AUTHENTICITY IN YOUR ONLINE INFO/ABOUT/PROFILE SECTIONS

When someone comes to your website, Facebook Page, LinkedIn profile for the first time, they're likely to read or select the "about" or "profile" or "info" section. Why? Because they want a human, a background, a connection and reassurance that you are real.

Here are some helpful tips when creating this piece of your online presence.

1. Don't use meaningless jargon. People want to get who you are without having to read something three times or look up words they don't understand. Remember, the goal of connecting is to not exclude anyone by sharing information that may have that effect. In addition, don't be a chest-thumper. No one wants to hear how great you are. Remember the goal is to not exclude anyone. Here is an example that, while impressive, may exclude a large segment of the population:

*.....is a recognized provider of result-based online and mobile advertising solutions. Dedicated to complete value chain optimization and maximization of ROI for its clients... is committed to the ongoing mastery of the latest online platforms - and to providing continuously enhanced aggregation and optimization options.*

I don't know about you, but when I read this, I glaze over. This information may appeal to a small targeted audience. If your business goal is to serve and connect with as many people in the community as possible, then communicate in a way that the majority of these community members will understand when sharing about you or your company. When sharing who you are and your background, create a story about you. Make it interesting and engaging and most importantly, keep it simple.

Here is an example of something simple and clear:

*We help you manifest good stuff.*

2.  When uploading a photo, use pictures of you and your team vs. stock photos. People really like to see and learn about you, and a picture is worth a thousand words. People can read and feel so much from the pictures you post, so make sure you are smiling, engaging, welcoming and warm. The more photos of you and your team, the better. The more "real", the better. People love pictures!

3. One of my favorite sayings second to "give to grow" is, "Make it easy for people to find you and learn about you". Give your contact information and make sure those links work. Don't tell someone to search for your company on Facebook. They won't. And if they do, they may have a frustrating experience. Is that how you want to start off your online relationship? Make it easy for them. Put your contact information with links to all the social networks you are on. Include the links on your signature on emails and all printed media. Save people time and make their life easier. One suggestion is to create a social media portal which is a landing page that contains all the icons with links to your online presence. You may have an icon for YouTube, Facebook, your website, your blog and LinkedIn. Now all your information is in ONE location. Here is my landing page, check it out: **shelleyroth.com**.

sroth@shelleyroth.com

The good news is, my name is not going to change, and people can land on this social media portal with links to all of the information about me. Have clickable icons for all the social media tools you use. Include an icon for your website, your email service provider, such as Constant Contact, and any other icons that will take me to places where I can learn more about you - simply and easily. **www.brandyoumadeeasy.com** provides a cost-effective way to create a Brand You page.

4.  Be who you are and that means REAL. Write like you talk and put your name on it. Tell a story about yourself, a true one, one that others will resonate with. My bunny story is told a lot and resonates with a lot of people. People buy from people they relate to – that's real-ationship marketing. It's that simple. What's hard is having the confidence to just be you.

5.  Include testimonials and reviews to establish credibility. Most social media tools give you the ability to gather this information from your connections. The best way to get references is to give them! Be sincere in who you give them to and make sure they are true. For every two or three you give, you just might get one back!

The next section of the book will delve into the top three tools and assist you with how best to use them to represent your virtual self in the highest, most real way.

## FACEBOOK, THE TOP SOCIAL NETWORK

I am going to start with Facebook and spend a lot of time on this section. With over 750 million people at the time of publishing, half of whom log on every day and spend 55 minutes per day, well, I would say your target market is on Facebook.

It's important to create a compelling Business Page (also called a Fan or Brand Page) for two reasons: 1) You want to engage your Fans and 2) You want to make it sticky so they want to come back for more.

It's becoming more difficult to maintain a thick line between friends and business when online with Facebook in particular since it is being used by all of these communities and groups. The absolute first thing you want to do on Facebook is to set your privacy settings. (If you don't know how to do some of the things I mention throughout this chapter, go to: **www.youtube.com/springboardw** and find the video for what you want to learn.)

It's simple enough to do by using the drop-down menu under Account at the top right of menu bar. Spend the time to review and set all items. You will be glad you did!

I would say that Facebook's digital body language is one of an open book when first using Facebook. Out of the box, just about EVERYTHING is turned on. Why would they do this and basically allow the entire Facebook community access to everyone? So they garner more "eyeballs", which equate to more users, which equate to more advertising exposure and thus more advertising dollars! So, make the time and review all of your settings. You will be surprised at what you learn.

**_Building a loyal Facebook Fan base is crucial to your success on Facebook_**

The most important activity of your Fan Page (business page) is to engage with your community.

Keep in mind that Facebook is a place to network and connect first. ALWAYS ask yourself, "What's in it for them?" You want to add value before you promote yourself. I like to use the 7 to 1 ratio of "value-add" to promotion. The ultimate goal is to have your Fans interacting with each other. Build a community. Be consistent. Be "sticky" – keep them coming back for more and "listening" to what you have to say.

What follows are some tips for putting your best face forward on Facebook. All of these tips are first and foremost about putting your best digital body language forward with the goal of better "REAL-ationship Marketing©" of you and your business. Included are best practices that I have used for myself and clients and advice I have gathered from many social media gurus out there.

1. **Personal Profile Page versus Fan Page**

    First, you don't want to use your personal account as your business. That's what your Fan Page is for.

Many people on Facebook become "Friends" with everyone who invites them. I don't suggest doing this. I am very deliberate about who I am Friends with on Facebook. I don't mind having them see pictures of me and my dog and my family and the parties I've attended. You can talk about your business and share links on your profile page, however, when someone invites you to be a "Friend", steer them over to your Fan Page by replying with a request to be a Fan of your brand and give them the link. Make it easy for them to like your page – give them the link (URL), and give them a reason to like your page (WIIFT).

**REAL TIP:** Grow your Fan base by importing your contacts from Gmail, Outlook, Hotmail, etc. The system will automatically find your contacts on Facebook and suggest you add them. This is a great way to grow organically.

2.  **Create your Business Fan Page**

You want to have a Business Page on Facebook. Why? It's free, and there are over 750 million people out there at the time of this book's writing who might want to interact and do business with you. It's easy to create a page. Go to **www.facebook.com/pages** and click the Create a Page button. The first thing you want to consider, before you even create your page, is the Page Name. This cannot be changed once you have over 100 LIKES, so you want to assure there are keywords in your name so when people search for you they know how to find you. Your Page Name is also used by search engines and can boost your ranking when someone is looking for what you have.

Your Fan Page photo stream is one of the few things you can customize on Facebook, so think marketing with photos. Remember, a picture is worth a thousand words, so put something engaging that will draw your Fans in. Remember to make it authentic! Have a CTA (call to action), and add a clickable link to each of the picture's descriptions.

The Info link is where you want to share information about your business. Lots of people will go to this section to learn about you, so make it brief and engaging. Use keywords in there as they are indexed by search engines. Include URLs to websites, Twitter, LinkedIn, etc.

**REAL TIP**: Include your Business/Fan Page on your personal profile page under Employment. It will be clickable and takes folks to your Business Page. Make it easy for people to click and find your business on Facebook and elsewhere!

3. **Choose your Applications wisely**

There are tens of thousands of apps on Facebook and outside of Facebook. Do your homework. Look at the reviews of the apps; see which of your friends are using them and ask them for input; read the reviews; look at the number of users. Remember, most all of the apps are from third parties, thus, you need to do research before using them.

Some of the popular applications include:

- Two applications for custom landing pages are **www.fb.com/iwipa** and **www.fb.com/tabsite**.

- Apps that allow you to post videos and pictures are popular on Facebook. Both are "Google juice". They are a great way to give to your users, and they will help you rank higher in search.

- The NetworkedBlogs app allows you to post your RSS feeds directly to your Fan Page wall.

- You must use a third party app like Fan Appz and Wildfire to do contests, sweepstakes and giveaways (**www.fb.com/promotions**).

- Constant Contact and Mail Chimp, email service providers, have opt-in email sign up apps for Facebook.

4. **Once you reach 25+ Fans, you can create your Vanity URL**

To be able to convert your nonsense URL into something fancy-looking, like "www.facebook.com/ mybusinesspage" you need to have at least 25 Fans. Once you do, go to **www.facebook.com/username** and click the "Set a username for your Page".

5. **Custom landing page**

If you are going to invest in one piece of your Fan Page, make it the Landing Page/Welcome Page where people arrive before they select the LIKE button. This is where you can welcome them to your world and ask them to become a Fan by clicking the LIKE button. You also can include an opt-in box to gather permission-based email as well as a video to welcome people. It can be as elaborate at you like, since it is customized. There are several sources of templates, so look around (**http://www.fb.com/applications**).

What happens when someone "LIKES" your page is that they will be notified every time you update your page's status. When someone likes or comments on your status update, this will be reflected in his/her profile as well. And when your status update gets a certain number of "LIKES" and comments, then it will be promoted, based on Facebook's algorithm, into the Top News section of a user's News Feed, so more people might see it.

**REAL TIP:** You are 40% more likely to be LIKED with a custom landing/welcome page.

6. **Grow your Fan base inside of Facebook**

Some steps to assure success:

- When starting to grow your Fan base, look no further than your Friends on your profile page. Some of these Friends could be potential customers or be able to promote your brand to potential customers. Once you have a significant amount of Friends, you may use the "Suggest to Friends" link to promote your page.

- Use Facebook directory (**www.fb.com/directory**) to discover other pages, groups, events and people that may benefit from your brand. You can search inside of Facebook and then join groups that make sense, where you can provide value to those groups. This is a great way to grow your Fan base. Just remember, it's about giving to grow. Add value to groups you join and pages you like. Communicate: Once you have Fans communicating on your page, make sure you always respond in a timely fashion to ALL comments. People want to be heard and they want to be responded to. So make it a habit to check your page at least once per day. Your goal is to make people love your brand and become active contributors and ultimately brand ambassadors.

**REAL TIP**: This is not a personality contest. Remember, it is not about how many Fans you have. It's about if they are STRATEGIC Fans.

- Consider that your page is like a really good bakery where people come back for more every morning.

  No one likes stale bread and donuts, so when you are managing your page, always keep it fresh and interesting for your Fans. They join your page to receive information and interesting stuff from you. Use pictures, links, videos, to keep your Fans engaged.

  Here are some things to consider so your digital body language doesn't lose you any Fans: Be aware of how many updates you post; don't use software that automates your updates and don't be repetitive.

  Keep your content fresh and your Fans will come back for more, and also be aware that your posts are streaming into their news feeds.

- Ask your Fans to post comments and LIKES. Engage them with questions and reply to their answers. The more comments and LIKES you have, the more likely you will show up in the Top News section of your Fans' personal profile pages.

- Use media content on your page. Ask Fans to post videos or pictures of your product or service and how they use it. Post videos of you and your team members in action; use video to answer questions your Fans ask.

- Use holiday themes, and go off-topic. You don't want to appear one-dimensional and always focused on your business product or service. So have fun and ask questions that have nothing to do with what you do. Make sure the answers you are requesting can be short. Remember that attention is hard to come by. Keep the answers one or two words or short phrases.

- Honor your Fans by giving them specials. Make the specials for your Fans only. Also acknowledge your "Super Fans". These are Fans who write on your wall and answer other people's questions and share tips as well. They also tag you on their Fan Page and speak out on your brand. Some Fan Pages use Fan Gating: When landing on a welcome page and clicking the LIKE button, another page appears and offers the new Fan something special – a top 10 list, a free consult, etc. This is called Fan Gating.

- You can send an update to Fans. Use this wisely and sparingly. You do not want to spam them. It should add value or provide something important to your Fans. What THEY think is important, not what you might think is important. Once again, always ask yourself, WIIFT - what's in it for them?  There is nothing worse than having Fans "tune you out" because you are overloading them with information.

- Ask you Fans for help. Everyone wants to help others. It's human nature. Just don't abuse this. You can ask your Friends to suggest your page to their Friends by clicking on the "Suggest to Friends" and Share links on your page.  You don't want to send the wrong message, so don't abuse this!

- Ask other pages to add your page as "Featured". When you do this, the logo of your page appears in the Featured Section of *their* page. People will see it and might click the link to go to your page. Do the same for those you want to share. Your goal is to build partnerships.

- Applications are plentiful on most social media platforms, Facebook in particular. You can use these applications to further promote your business and yourself, however, research them to assure reliability and usability since they are from third parties, not Facebook. Perception is reality, so be aware of playing games that may give certain users a perception of who you are. Don't overlook that there are cases of people creating relationships from game-playing that ultimately lead to business. Just consider that others may judge you digitally by the company you keep. As I have stated before, it's best to keep your online persona in the neutral zone so as not to exclude others from relating to you and buying from you.

- Check out your competition and what they are up to. Learn from what you like and what you don't like about their virtual body language. No need to reinvent the wheel!

- Advertising is available to promote your brand. You can find out a lot about the numbers out there in your target market by selecting key demographic information and establishing your market before purchasing click-through or impression advertising. Check it out.

**Facebook User Attitudes on Ads:**
*What do you think of Facebook Ads?*

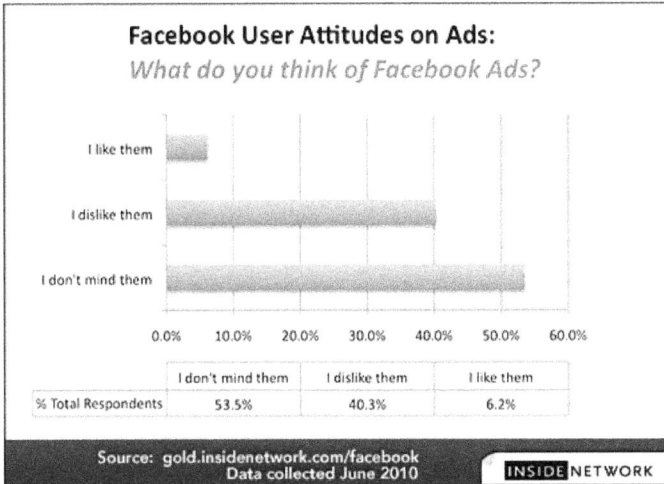

| | I don't mind them | I dislike them | I like them |
|---|---|---|---|
| % Total Respondents | 53.5% | 40.3% | 6.2% |

Source: gold.insidenetwork.com/facebook
Data collected June 2010    INSIDE NETWORK

Other ways to promote your Fan Page outside of Facebook include adding social plug-ins (buttons to access your brand) to your website, blog, other social media accounts and RSS feed. In addition, you can add the Facebook Like, Share and Send buttons on any of the sources above. This ease of access makes your website more personalized for the user experience. Friends are more likely to like a brand and buy that brand if they see their friends doing the same. Thus, the Like, Share and Send buttons are being seen more and more. Having those buttons on your website and blog will drive more visitors. Check out **www.fb.com/developers** to learn how to add these social plug-ins.

7. **Include links (URLs)**

   Make it easy for people to find you and learn about you. Put all of your URLs on letterhead, business cards, email signatures, LinkedIn profile, Facebook profile, Twitter info and blog info.

   **REAL TIP:** When typing a message into your status update, make sure you use the @ symbol to tag other business pages, groups and people you want to give a shout-out to, including your own business, when appropriate. Just start typing the name of the page you want to mention after the @symbol.

8. **Use Video to promote your links**

   Create videos. They can be instructional, funny, presentations and anything else that may engage your target market. Then make sure to include links to your social media pages and profiles with a request to have them join.

   **REAL TIP:** People love to laugh. Just because a video has nothing to do with your business doesn't mean you won't be highly regarded if you share an appropriate humorous video every now and then.

9. **Include all your contact information on all of the media you use.** It's about business and making it easy for people to find you.

10. **Don't forget to watch your analytics so you can see what is working.**

Analytics are free at **www.google.com/analytics**, as well as Insights on your Facebook Business Page. Another great source is **www.twentyfeet.com**. Use them!

### Footnote on Privacy

There is a lot of press on privacy and the lack thereof on these social media tools. I am going to just say four things about privacy:

- If you don't want it public, don't post it.

- Take the time to explore and set ALL your privacy settings in all the networks.

- Don't be an open book. Consider keeping some of your life private. Even though we are all more connected than ever, no one needs or wants to hear about your every move in every waking minute of your day. BE STRATEGIC!

- Don't announce when you are not home. Do you really want the bad guys to know that?

**REPEAT: IF YOU DON'T WANT IT PUBLIC, DON'T POST IT!**

### Additional ways to ENGAGE your Fans on FACEBOOK

- When someone becomes a Fan, send them a message thanking them for becoming a Fan of your page and give them a call to action, such as "stop by often and ask any questions you might have on x".

**REAL TIP**: Monitor your Fans. The way to know the newest Fan is to look at your list. The last picture icon is the last person that became a Fan.

- Continue asking for some type of call to action. Use questions regularly to promote interaction between you, your Fans and each other.

- Share links to third-party articles and ask your Fans what they think by asking a question about the article.

- Be strategic in your use of tags (@ symbol). You don't want your digital body language to "say" you are too loud or beating your own drum. (This also applies to tagging other pages and Fans.)

- Don't bury your head in the sand when it comes to someone sharing comments that are negative. Address it. Some of the best marketing campaigns I have seen come from turning a negative situation around via facing it head on.

- Encourage others to LIKE your updates. It's ok to LIKE and comment on updates yourself as they are posted.

- Share by tagging other people and their businesses and by posting other people's links.

  **REAL TIP**: Use your people skills and be empathetic. Strive first for social success and NOT marketing success. BE REAL! "REAL-ationship Marketing©" works.

*Ways to make sure you're being real when creating your Fan Page*

You want your Fans to get you and relate to you and each other. It's about engagement and building trust. Here are some added ideas for creating and maintaining your Fan Page:

- **Ask yourself, "What is the purpose for my Fan Page?"** Interestingly enough, the number one answer is **not** to sell more "stuff". The top objectives include building your email list, increasing traffic to your website or other social sites, promoting events and registrations, branding your company, connecting with affiliates and partners, establishing yourself as an authority and providing exceptional customer service.

- **Choose your Fan Page name and category carefully.** They cannot be changed after your Page receives 100 LIKES. Use keywords in your page name for search engines to find you.

- **Add a custom Welcome tab to your Fan Page.** (This is where non-Fans will land before becoming a Fan.) A recent study has shown that only 25% of people will LIKE your page if they first land on your Facebook wall versus a custom Welcome or landing page. Check out third-party application templates for creating custom landing pages (**www.fb.com/applications**).

- **Post about you and your products or services minimally.** Remember, ask yourself, what's in it for them? Provide valuable reasons for them to read your posts. Include blog posts, articles, podcasts, videos, photos, OPC (other peoples' content). Your goal is to provide value to your Fans by making it a one-stop shop for finding a variety of useful information in your area of specialty. **www.technorati.com** compiles links to the top blogs, so it can be a good source for OPC to post in your social media. Another source of OPC is **www.alltop.com/your topic** (e.g., if you're interested in photography, search **www.alltop.com/photography**.)

- **Always allow Fans to post on you wall.** You want to be engaging, so encourage sharing of videos, photos, links. If someone posts something you do not like, just remove it. Always reply to your Fan's comments with comments of your own. Engage!

- **Keep your page fresh**. Provide updates, at least one, every day. If you have something valuable to share more often, that is fine. If you build it, you must encourage people to come. Maintaining your page takes time and thought.

- **Use your page for calls to action:** Contests, Fan-only offers, give-aways, questions that engage. All of these will keep your page sticky. (Remember: You must use a third-party app for contests, promotions, etc. Check out **www.wildfireapp.com**.)

  **REAL TIP:** Before you create your Fan Page, list your objectives for having one in the first place.

Keep in mind that Facebook Pages are searched by all search engines and can give you and your business even more exposure than your website.

In conclusion, people who connect with you on your Facebook Fan Page are interested in what you have to say and offer to them. Be real with them and you will create raving Fans that advocate your business to their friends and associates. The power of Facebook is in the numbers. I can't think of a better compliment to who you are being on Facebook than having someone share your page, your content and who you are with their connections.

So always put your best FACE forward!

Let this guide be a starting point. Get creative, use your imagination, invent new ideas of giving to grow. There are lots of ways to use social media marketing to identify and grow your target market. Just remember to be real!

## LINKEDIN, THE PROFESSIONALS' NETWORK

As THE professionals network, it is most important to consider your digital body language when developing content for this social media platform. It's like your mama used to say, "You have only one chance to make a first impression."

LinkedIn is used by professionals. Human Resource departments are always searching LinkedIn to learn about potential employees, so it's really important here to put your best professional self out there. Business-to-Business users and those looking for jobs frequent LinkedIn for various reasons. You can establish yourself as a brand expert by answering and asking questions that position you to share your knowledge. LinkedIn is an amazing networking tool to meet just about anyone you want to meet. It basically lets you connect with people that are three degrees from you. So, if I know Brittney, and Brittney knows Mike, and Mike knows Rob, I basically can meet all of Brittney's, Mike's and Rob's connections just by asking for an introduction.

**REAL TIP:** When sending out an invitation to connect, always personalize the note you send. It shows thoughtfulness, caring and personalization... and takes 10 seconds.

**Picture:** As we've said in previous chapters, one of your single most important digital body language indicators is the picture you post on all social media sites. Your picture on LinkedIn is most important. You don't want to put a picture of your dog, your children, you golfing, etc. A professional head shot is what you want, and you want it to be current. For the longest time I had a picture of myself that was 10 years old. Not good. It should be a high-quality, professional image up there for all to see. Don't go "public" till you are ready for people to see your profile. You can turn off notifications while you are working on getting your profile page edited (**www.linkedin.com/settings**). Every time you make a change, updates are set to alert your connections that you have made a change, UNLESS you have taken the time to turn this setting off. I suggest keeping this off until you are ready to go public. (Just go to Account Settings to turn it off.)

**Profile:** Your profile page is what people will land on when they look you up on LinkedIn (**www.linkedin.com/ in/shelleyroth**). You want to capture what you do, with some key words sprinkled in. We are all multifaceted, so make your profile interesting and intersperse some information about who you are, not just in your job. For example, you may be the coach of a Little League team. Tell a story about yourself, and keep it in the first person when writing your profile summary page. Add your skills and specialties to your profile (**www.linkedin.com/skills**).

**Groups:** Groups you join or start are important vehicles for sharing information and insight (**www.linkedin.com/ groups**). You are allowed 50 groups, which can accumulate quickly. Make sure the groups you join give the right impression and reflect positively on your digital body language. Whether it's business or personal, group icons that you belong to will appear on your profile page, and I know I look at which groups people are in to assist in formulating my impression of their digital body language.

In addition, your target market may be involved with various groups on LinkedIn. Be conscious of what you represent when joining groups and answering questions. You do not want to be viewed as a lurker in a group. If you join a group, do so to participate and add value. One of my pet peeves is people joining groups just so they can sell you something inside the group. Taking that a step further, they try to sell you subliminally. One example of this was a group I joined around my area of expertise, social media. The group "owner" grew to thousands of members. Then, he used the group for his own self-serving purpose - to sell. I am sure I wasn't the only one in that group who felt betrayed by this "leader". Be very aware of who you are being out there. As stated before, the easiest person to be is yourself. No hidden agendas.

Another example: In one of the groups I joined, a very bright friend/business associate of mine did something so obviously, blatantly surprising to me, that I had to address it with her. She joined a group, and then posted an event she had going on. It reminded me of watching a show and then an advertisement comes on, having nothing to do with the program I was watching/listening to. I immediately wanted to tune her out, and her "stock" dropped for me. I gave her the benefit of the doubt and talked with her about bringing value to a LinkedIn group vs. selling to the group. She was all about what's in it for her vs. WIIFThem... Huge difference. Had she joined the group and then given some valuable lessons, tips, etc. and maybe after becoming an active value-added participant told others of the benefit of her workshop, well, that would have been socially, digitally, acceptable. Her digital footprint wouldn't have left a bad taste. Now I am wary of what she has to say/sell. The SPIRIT of groups on LinkedIn is to share and grow, not to sell!

**Q & A:** Question and Answer on LinkedIn is a great way to make an impression and establish yourself as an expert when asking and answering questions (**www.linkedin.com/answers**). Be aware that there is a right and a wrong way to do this. Use the tips on LinkedIn to assure you are not perceived as trying to jockey for position.

I asked a question on LinkedIn, and it had a component of a legal issue. Well, there were at least 12 lawyers lying in the weeds waiting to pounce on questions related to law. Now I will tell you that 80% sincerely wanted to help; however, the other 20% were just in it to convince me to hire them to do the work. People will read your digital body language inside of the Q & A section, so ask and answer questions sincerely and remember, always come from "give to grow".

Most importantly, be aware of the pulse inside LinkedIn questions, answers and groups. Study other LinkedIn users and how they respond.

**Here are some ideas on how to grow your connections on LinkedIn organically.**

1. **LinkedIn is all about connecting with people who are two or three degrees from you.** That is the power of this social media tool. Don't be afraid to ask your number one connection for an introduction to someone in their network. (They will follow through or may choose not to.)

2. **After meeting new people in live venues, take the time every week to invite them to connect to you on LinkedIn.** Don't forget to make your invitation to connect personal. Keep the conversation going.

3. **Before you attend a meeting with someone, look up their LinkedIn profile, and invite them to connect with you.** This will give you information that will be valuable in the meeting and show the person that you care. It's all about building relationships.

4. **Before you consider removing someone who is sending you sales messages from your network, be aware of who they are networked with.** You may not want to lose that second and third degree of potential connections.

5. **Be aware that some people just aren't real.** They may have a profile created that is false or contains false information, just to have that juicy worm on the hook. Do your due diligence on everyone, and decide for yourself what their digital body language says to you. 90% of the time, I will not connect with someone if they do not have a picture of themselves since it's a red flag for me. (I ask them if they need help uploading a picture to try to help get at why they don't have a photo up.)

6. **Upload your contact list via webmail, Outlook or .csv file,** and LinkedIn will let you know who is already on LinkedIn. Then you can invite them to connect with you.

7. **Check your alumni groups and past employer groups.** This is a great way to connect.

## TWITTER, THE REAL-TIME MICRO BLOG

Twitter is a micro blog (**www.twitter.com**). You have 140 characters to use for the messages (tweets) you send. The messages go to your "followers" and in turn, you will receive messages from those you "follow". You have a very limited space to share information about yourself, so use the background area wisely. Include your website link and a short "bumper sticker" bio about what you do. Your background graphic can be customized or you can use one of the many templates out there. Twitter is a fantastic communication platform and real-time search tool.

Your intention on Twitter is important to your digital body language. This is a platform where you want to "listen" to the conversations occurring before you jump right in. There are certain rules to follow which we will cover here. The best rule is to observe first before you send your first tweet.

**Here are some tips for using Twitter effectively adapted from Kerry O'Malley with Marketects (www.MarketectsInc.com):**

1. If someone follows you, follow back, unless they are spammers. I always look at someone's Twitter page before I follow back. There are a lot of self-promoters out there, and for me, it's about strategic relationships and not just numbers.

2. A good rule of thumb is to promote others and share value 12 times to 1 self-promotional tweet.

3. Retweet (RT) information that you find valuable, and always give credit to who you got it from by using the @ symbol in front or their handle (name on Twitter).

4. Many people use auto-tweets. I think they are less caring and human, so I am not a big fan of what this digital body language says about someone.

5. Do not self-promote or spam. You will be unfollowed. Use Twitter as it was intended, as a personalized communication tool, not as a way of selling.

Let's discuss Twitter as a marketing tool and some of the rules to follow. There are generally three main categories of users on Twitter. "Relationship Builders" use Twitter as a communication tool, "Promoters" self-promote via this micro blog and "Power Users" have tens of thousands of followers and are constantly tweeting.

These three types of users may not see eye-to-eye on how Twitter should be used. You may view these types as self-serving and only out to make a profit or as arrogant blowhards. It's crucial for you to be who you are as it is a very transparent communication network.

Here is a bit more of a description on the digital body language you might see on Twitter based on the categories described above.

**Relationship Builders** almost always follow people who follow them. They try hard to insert @ tags which acknowledge others and their tweets. They will send out DMs (direct messages) and expect to get a response.

**Promoters** rarely post anything of a personal nature. Almost every tweet is about them and what they do for a living or what their company does. They really don't communicate via DM (direct message) unless it brings in a potential customer. They won't use the @ tag unless they are trying to promote their own agenda.

**Power Users** usually have many more followers than people they follow. They post often. They rarely use the @ tag or DMs. It's all about them. Celebrities or business gurus most often fall into this category.

The bottom line on Twitter is to be very aware of your digital body language and the message you are sending to others. As on all of the social media tools, there is the potential for being misinterpreted and thus misunderstood.

Here are some tips for being REAL with your Digital body language in the Twitterverse:

1. **Complete your profile information and use your real name**. I didn't look right or left when I created my "handle" springboardw, and now I realize this was a mistake. People talk to people on Twitter so use your name. Upload a photo of you, not your

dog, your partner or your company logo. (I like to see you better.)

2. **You should have a similar ratio of people you follow** and those that follow back. If you follow 10 times the number of people than follow you, you may be perceived as a spammer. I always look at this ratio when deciding if I want to follow back.

3. Twitter is a public platform so **if you don't want it public, don't post it!** All posts are permanent and will tag to you. Remember that the search engines like Google pick up the conversations by you.

4. **Don't over self-promote**. Once people follow you, they will know what you do. You do not have to remind them of this. If you self-promote, you will lose followers. Not many people like chest-thumpers.

5. **Use DM (Direct Messages) for ongoing private conversations** with someone. No need to take a discussion out there for everyone to read. Twitter can be very "noisy" and this cuts down of some of that noise.

6. **Respond to @ and DMs.** This is just common courtesy. If someone asks you a question or comments to you directly, respond to them in a timely fashion. This will say a lot about who you are being in the digital world and definitely gives an

indication of your digital body language. Just as on Facebook and LinkedIn, monitor all comments daily.

7. **What would Miss Manners say?** One of my colleagues always asks, "What would Snoopy do?" Well, in this case, think about basic manners and applying them. Be real, however, be respectful of others, and don't exclude others by voicing opinions that might land you in hot water.

8. **Don't Tweet when inebriated!** This actually applies to all the social media tools. Use your head. You don't want to send messages when you are coming from an emotional place. Remember the saying, "loose lips sink ships". Well, the same applies online. Step away from that computer or phone when partying.

Twitter is a great tool for engaging and building relationships and sharing valuable information with your followers. To learn the Twitter basics, go to **www.shelleyroth.com/store**.

## EMAIL MARKETING TIPS

Email marketing services are a great opt-in tool for customer and prospect communications. Here are a few words on making sure your information gets through to those who want it.

Most of us are already deluged with more information than we can handle; entering the social media world expands this reality even more, so many people are using email filters to manage their incoming information.

By design, filters try to decide which emails you want and which ones you don't want in your inbox. These automated guardians can be strict, lax or even unpredictable, but they all work via a mixture of generalizations and hard rules.

*The issue is that sometimes they can inadvertently block out information like yours that your community has opted in to receive.*

Since your customers have opted in to receive information from you, you know they want this email. The question is how to make sure that the filter knows it should be placed in the inbox as opposed to a spam/bulk folder. Constant Contact has a built-in spam checker that will scan both your email and subject line to look for content that might raise some flags. This is a very valuable tool that can highlight any number of trouble areas for you.

However, being aware of what you are typing and how it may be seen by an automated filter is half the battle. Using the words "Free," "Guarantee," or "Credit Card" with any frequency in the email or subject line is a rock solid no-no.

You should also avoid excessive use of punctuation, capitalization or symbols. While "Amazing Deal!!!" or "CAN YOU AFFORD NOT TO?!" may seem energetic and magnetizing, automated filters don't pick up on the emotion, just the cold, hard punctuation. A single symbol is safer.

There are, however, people who use filters with customized rules. So if you want to cut to the heart of the matter, getting your email subscribers involved can have dramatic results. Asking your subscribers to add your address to their address book or even to their "Safe Sender" list helps it become even clearer that your emails are wanted.

As a last thought, authentication is an option that is available for every account and has both immediate and long-term benefits for your deliverability.

In the short term, it reassures that the email is not a spoof and that your email service provider is sending it with your permission, on your behalf.

In the long term, it helps establish your reputation as a trusted email marketer and that translates to your emails getting into the inbox more often overall.

In conclusion, I hope this section has given you a good overview of what these tools are all about and how they can impact how you are being "seen" online. Invest the time to learn by observing first before taking action. See what feels good to you and who makes you want to come back for more. Trust your instincts when you are evaluating people and their companies on social media platforms.

If your gut tells you something isn't right, listen to yourself! Steer clear. And be YOU!

## Chapter 9
## REAL-O-METER: How you communicate online and ensuring your words reflect the real you

The REAL-O-METER is under development through the use of various algorithms, tags and keywords. You will be able to use this web-based app as a fun experiment for entertainment purposes to try to assess the "realness" of your and other people's social media entries like pages, blogs, profiles, brochures, etc.

We will be continuously updating and refining this tool as more information is gathered.

### BE AWARE! WHAT YOU SAY CAN BE INTERPRETED IN MANY DIFFERENT WAYS

*How we do anything online…When we answer a post or comment on other posts… Interacting with people we know and people we don't know…*We are never sure how we will be interpreted online.

I've already shared that I know I can be perceived of as thoughtless; I don't mean that negatively, just that we respond without thinking about how our communication is interpreted. Sometimes we are just reacting vs. proactively thinking about how our communication lands on someone. When we are in front of people, we might be more sensitive to our words since we can see their body language and determine how they are reacting. This allows us to clean up, adjust or clarify something that didn't land as intended. When we are online, it's just the printed word with no physical body language to help us interpret. Granted, caps, colors and emoticons can give meaning to an exchange; however, basically, it's text representing you.

I can assure you that everyone interprets what we say online differently. You can ask the same question of 10 people and get 10 different answers. If you say something, and it lands on someone the wrong way, you are going to alienate that person, customer or prospect. You want to be very thoughtful about what you are saying. Does the person you are communicating with have a "listening" for you, and what you are communicating? If you are unsure, clarify that they understood your intended meaning.

How do you do that? Ask them! "Did this answer your question?" "Does this make sense?" "Is this what you asked?" This is very important when you are communicating with your community of Fans, followers and connections on social media. Always qualify that they understood and you answered the questions. Any thoughtful salesperson tries to qualify, not that this is primarily about sales.

When you answer someone's comment or respond to a blog post, and you are assuming you understand what they meant in their communication, I assure you that the likelihood is high that you will hear back, "No, that is not what I meant."

Unless a communication is black and white, it can be misinterpreted, especially when in text format where you can't read the person's physical body language.

## HOW WE EXPRESS AND DETECT REAL [AND UNREAL!] WITH WORDS AND PHRASES

Words, communication, tone of voice, how you sound in-person and online are so important. Are you engaging? Are you listening to TONE? The third cue in effective communication is visual, but is there a way to read visuals online?

Are you being real, or are you being someone else? Are you being what your boss wants, your parents want, your spouse wants, expects, reflects?

In creating the REAL-O-METER, we aren't applying rocket science: A team of people helped me subjectively generate a list of words and phrases. One set were "potential positives". Another set were "potential negatives". Then we weighted them – also subjectively – from 0 to 100 on a scale of realness. Low numbers mean less real and high numbers mean super-real.

The REAL-O-METER app will search out the social media entries you select to apply it to [like a blog posting or a Facebook profile] and will note the frequency of any of the words or phrases in the REAL-O-METER Index. After doing some calculations with the frequency and realness weightings, it will return back an assessment on the REAL-O-METER dial graphic, noting whether that entry is:

- **"Hazy, Try Again"**: meaning there weren't enough words or phrases in the entry to give it a rating
- **"Our Sources Say, Get Real"**: meaning it received a very low realness rating of 1-25%
- **"On the Right Track"**: meaning it received an average realness rating of 26-50%
- **"Outlook Promising"**: meaning it received a higher realness rating of 51-75%
- **"All Signs Point to Real"**: meaning it received the highest realness rating of 76-100%

In the spirit of transparency and realness, I'd like to share a sampling of our most recognizable highest ranking "potential positives" words and phrases and some of our lowest ranking words and phrases of "potential negatives".

**HIGHEST REAL-O-METER POTENTIAL POSITIVES**

Benevolence
Civil society
Common good
Compassion
Conscience
Conscious business
Conscious citizen
Conscious living
Diversity
Divine
Doing good
Empathy
Environmentally responsible
Fairness

Human dignity
Inclusiveness
Inner truth
Justice
Living consciously
Making a difference
Right relations
Service-oriented
Social good
Soul-centered
Spirit of goodwill
Unity
Universality
Wisdom

Goodwill (good will)          Wise use of power
Harmlessness                  World server
Higher nature                 World service

**LOWEST REAL-O-METER POTENTIAL NEGATIVES**

Blowout sale                  Free selling kit
Chance of a lifetime          Going out of business
Don't be left behind          Miraculous results
Don't delay                   Last chance
Everything must go            Today only

Even if you don't use the REAL-O-METER, the words and phrases above can be a roadmap of what rings true and what strikes false notes. Have fun experimenting with your natural communication style, and see if you tend to use more positives or negatives in the way you interact.

Negatives aren't just about the phrases we provided above. If you have a generally negative attitude that diminishes or criticizes yourself or others, it's a good bet people won't want to be around you in social media. We don't usually like to be in this vibe in person, so why would we want to digitally? Being real online just takes common sense and using good communication practices that you're hopefully using every day "out in the real world". If you aren't, refer to the Digital Body Language Muscle-Builders in earlier sections of this book to develop those skills further so you can let the real you shine through – anywhere, anytime!

**IF IT'S ON THE NET, IT MUST BE TRUE: HOAXES**

Here's an interesting article from CrunchPress.com about online hoaxes.

> "This past week, formerly unknown actress Elyse Porterfield fooled millions playing Jenny, the fired Dry Erase girl, in a clever hoax. Right now, I guarantee other pranksters are dreaming up new schemes to fool you again. And journalists, who at one time were tasked with protecting the public from such lies, no longer have the same power to block them.
>
> The media has reporters and editors in place to prevent hoaxes from going public. Sometimes it works, sometimes it doesn't.
>
> The Jenny story was a fun, harmless, light story. But, hoaxes can involve a story of major global importance. I was in the CNN newsroom in 1992 when our medical reporter got exclusive information that then President Bush had died at a dinner in Japan. The first reaction was shock. Then, the newsroom got very quiet and we struggled to decide what to do. Trust our reporter and break the story or get a second source and risk someone else beating us on the story? If we decided to go with it and we were right, CNN would have become the first to tell the world this huge news and gain credibility. [If] we reported it and were wrong, we would

*have spread a huge lie and suffered a major embarrassment.*

*CNN decided to get confirmation first and soon discovered it was a hoax. But our sister network, Headline News, started to report on the "tragic" news until a wise producer told the anchor to stop.*

*For the most part, we were able to prevent the hoax from getting out. But, today it might be different. Just one unauthorized tweet from inside CNN could start an unstoppable viral explosion. ABC's Terry Moran created a stir last year by tweeting about an off-the-record comment where President Obama called Kanye West a jackass in an interview that had not yet been edited for air. Certainly not a story of major importance, but it does suggest that media organizations can't control their employee's tweets.*

*I always hated working on the early morning shift at CNN on April Fool's day because I knew we'd have to vet some hoaxes at 4 AM. Our credibility would suffer if we presented fake stories to the world. One year, Taco Bell bought ads in major newspapers saying it was buying the Liberty Bell and renaming it the Taco Liberty Bell. We reported on the ads, but that it was also April Fool's day. AOL ran a headline that life was discovered on Jupiter. That one generated 1,300 messages on AOL (huge for those days) and was considered by some to have risked AOL's credibility.*

*There were many other fake stories the public never heard because they were vetted by journalists. And without the media, there was no way for these stories to spread.*

*That's all changed. With social media, there are no editors. There is no waiting for confirmation. When you tweet or re-tweet, you are not checking the facts or even so much concerned if you are spreading a lie. When the Dry Erase Girl meme hit the Web, 421,000 users shared the story on Facebook, and theChive got 2.5 million unique visits for two days in a row, the same amount it normally gets in a month.*

*Jenny's story, which many people wanted to believe, was posted on theChive at 4:30 AM. The site claimed they got the photos from someone who works with Jenny, but they even admitted Jenny's name was not confirmed. It was then re-posted at College Humor, and by many others including TechCrunch, on which Jenny had claimed her boss spent 5.3 hours a week reading. Really just bait for us to run the story.*

*TechCrunch's first post included a link saying "via The Chive." It was a water-cooler story about a spreading internet meme with the playful headline – It's Official: The Best Bosses*

*Read Techcrunch." It was an amusing slideshow, regardless of whether it was true or not.*

*At the same time, our reporters were trying to contact Jenny. Were there some red flags? Sure, Jenny, if that was her real name, had no last name and the name of her company wasn't mentioned. But, this is how process journalism now works. It's journalism as beta.*

*In the days before social media, I think news organizations might have held the story. Now with the instant viral spread of information that happens even without the media, the story is out there whether we report it or not. The man behind the hoax, theChive's John Resig told TechCrunch "we didn't need mainstream media to make this happen." We certainly helped to spread it (and debunk it) faster than it would have on its own. But did TechCrunch and all the other media who published the fake story suffer a loss of credibility? You tell us. But my sense is, not really. We helped move the story towards the truth.*

*Peter Kafla at All Things Digital was the first to publish some skepticism, saying the story was "almost certainly made up." TechCrunch was first to confirm the true identify of Jenny that night and followed with a video interview the next day. Yes, we extended Jenny's 15 minutes of fame and it was also our most popular video of the day. Jenny told us she just taped an interview with Edward R Murrow's CBS News before our interview. Even if a news organization didn't report the initial hoax, many will do the story behind the story and what, if anything, it means. Guilty here, too."*

## A PICTURE SPEAKS A THOUSAND WORDS, BUT WHAT WORDS ARE THEY SAYING: HOW WE DIFFER IN INTERPRETING

Here's a great example from comments on a blog post by Mike Rodgers of how something even as simple – and you would think uncontroversial – as smiling can create totally different impressions, interpretations and opinions.

### 8.5.10 Picture speaks a thousand words: Does Smiling Make you a Better Leader?

*How many of you smile as much as you could? Here are four reasons everyone, including and especially those that lead people ought to smile.*

*1. Smiles make you happy. If you are ever in a down mood, try smiling ear to ear. Watch your mood change instantly.*

*2. Smiles make others happy. People like to be around people that are happy, it makes them happy.*

*3. Smiles say you care and are approachable. Research shows that being kind to employees for example improves productivity.*

*4. Smiles use fewer muscles than frowning. Therefore smiling will make you less tired and give you more energy.*

*One thing to note - make sure when you smile that it is genuine and not forced or fake. People can tell the difference. So smile today. Smile now, feel the difference. I am sure this young operator had no idea the difference she had on me that day in the amusement park. What impact are you having?*

*Can you think of other reasons to smile? I welcome your comments. Thanks.*

*COMMENTS:*

*smile is contagious, it adds charm to one's personality--A leader has necessarily to be smiling .His smile will make the aura all around motivating and will make the workplace a productive place*

*Posted by: saiyid | 08/04/2010 at 10:04 AM*

*Well, for one, a smile or laugh can cover up alot, good and bad wether you are delivering the smile/laugh or receiving it.*

*A smile means alot.*

*TDK of TDKtalksTM.com*

*Posted by: TDK talks TM | 08/04/2010 at 10:34 AM*

*Nice blog article Mike. I agree - nothing beats a real genuine smile. It's an ice breaker in any situation.*

*Posted by: Nancy (Range) Anderson | 08/04/2010 at 10:38 AM*

*I like the Chinese proverb that 'man without a smiling face should not open shop'. Same thing for business.*

*Posted by: Paul Boross | 08/04/2010 at 10:39 AM*

*I tend to disagree that smiling make you a better leader. It may make you appear more approachable, but leadership is not about being approachable, liked, or happy. There are many smiling, happy-go-lucky, cheerful, people who when it comes to leading others are total screw-ups of whom it would be foolish to follow.*

*Leadership is about taking the point position. It is about forging the path, finding the best route, guiding others, being the person to risk being wrong, being the first to volunteer, to champion a cause, to both receive and give direction. Does*

*smiling enhance your ability to do these things.
No.*

*However smiling can influence others behavior.
A smile doesn't alter Your LEADERSHIP abilities.
However, it might alter Your INFLUENCING
abilities.*

*Influence is often considered an attribute of
leadership in general, but it is an attribute of
both good and bad leaders. I'm sure Satan
smiles when he influences us to do wrong.*

*Posted by: John Walton | 08/04/2010 at 01:03
PM*

*Many leaders can take themselves too seriously.
While leadership may require one to be tough
on others a smile can let others know you have
a balance with your life and maybe theirs. Life
can be too serious to be serious at times.*

*Posted by: Stephen Kendrick | 08/04/2010 at
08:40 PM*

*Thanks for your comments John. I agree with
some of what you have said and don't with
other things. Smiling can make you a better
leader. I would agree that smiling by itself
doesn't make you a great leader any more than
"being the person to risk being wrong" (as you
state) or "being the first to volunteer" (as you
state) alone makes you a better leader. It is a
component of being a good leader.*

*Being able to smile says you care and are approachable. I disagree with you that this isn't important in leadership. Leadership is about people following you. It is about inspiring people. It's about giving them a vision. It's about influencing. People follow people they trust (I understand that people can be led by dictators as well). People trust people they know care about them. Again, smiling helps you appear more approachable and caring, which increases your trust, which increases your ability to influence and have people lead.*

*There is good leadership and bad leadership, but both are leadership, right? Hitler was a leader, but he led by influencing through fear and lies. Jesus Christ was a leader, but he chose to lead and influence by pure love. Both smiled, but one was genuine, the other was evil. I am certain that when Jesus smiled it created feelings that he cares.*

So, the REAL-O-METER's goal is to help make us aware of what is truly heart-felt vs. self-promotional in our communications.

When you are online - because it is such a two-dimensional communication - the first thing you want to do is always ask yourself, are you coming from your heart, or are you concerned with getting vs. giving? Once you practice just being from "give to grow", it becomes more natural, and you have to question and review yourself less often before you hit the send button.

While you're in practice mode, the list of words that we've included as positives are a guide to help you back to "give to grow" if you find yourself slipping into the old habit of "give to get". If you do feel yourself slipping, just think about incorporating positive words and phrases from the REAL-O-METER into your message. It will be less negatively received in our opinion by those who are reading your digital body language.

If anyone with technical software development and programming skills resonates with the idea of the REAL-O-METER application and would like to work with us on development, please contact us at sroth@shelleyroth.com.

**Chapter 10**
**Beyond the digital frontier: Getting real in the real world**

I have asked myself many times, why did I write this book?

What started me on this path was stumbling on that wolf in sheep's clothing as I started my education on social media. What I have shared throughout this book are my life experiences that brought me to this stage of my life. What my energy advisors helped me see is that every single person has a purpose, and my purpose on this Earth is to be REAL and help others be real.

So, I want to leave you with one last tip. Follow your heart. Turn off the noise, and LISTEN to your intuition. If you have to go off on a vision quest, do this. If it is in the form of taking 15 minutes a day to be still, then be quiet. We are all here for a reason. Every single one of us. You may not have discovered that yet, but it has discovered you and is just waiting for you to access it.

We as human beings are constantly in a state of doing...Human Doings instead of Human Beings...always busy and active, mostly ignoring our calling to a higher self. We have exercised our FEAR muscles for so long that our VISION muscles are total flabby! Just like our physical muscles, our VISION muscles need to be dusted off and addressed on a regular basis so we stay in touch with our goals, dreams, hopes and aspirations for who we truly are. When we see that we can live our future today vs. someday, tomorrow or next week, we begin to exercise our vision for ourselves, and anything becomes possible. It always takes a lot of work and a wonderfully supportive network of friends, loved ones and great advisors or coaches to keep us on track. But now we have new additional tools to help us with this process: social media.

Your virtual body language is an opportunity to take the essence of who you are to the entire planet. That's the beauty of the online world. You can be in front of the masses and share your gifts with hundreds and thousands of people. They will "hear" you if you are true to yourself and listen to your heart. There are many people who will tell you what to do and how to do it. You may think they are smarter than you and wield more influence than you. Trust me, there is only one you, and I believe we are all guided to do the right thing if we are aware of our purpose and if we are just still enough to listen to our own intuition, our gut, our heart.

We were put on the earth for a reason. Discover that reason and make it happen.

Stop listening to others and trust your own self to be guided to do the right thing.

Go forth and PROSPER! As Gandhi said, "Be the change you want to see in the world!"

So my fellow social media travelers, I hope I have helped you navigate through some of the social media waters out there and also encouraged you to take off the masks that we wear every day, get rid of the avatar that's not you and let your true real self come on out to play! You cannot lose when you give to grow and .......remember always to.........GET REAL, PEOPLE!

**ULTIMATE REAL TIP:** Stop, look and listen to your physical body. Trust in your higher self and your innate knowledge of what is right vs. wrong, what is real vs. false, what is give vs. get. Get to know the landscape of your heart, and be creative about how you express it through this amazing new world of digital social media communication. (See Appendix I for a master list of all the links we've provided in this book to connect to the world of social media.)

*To continue your exploration of getting real online, visit on.fb.me/engagecustomers to receive your free e-copy of "Engaging Customers with Social Media".*

## ACKNOWLEDGEMENTS

This book is dedicated to the many people I have encountered over the course of my blessed life. Starting with the authors and teachers that influenced who I have become, I would like to thank Carolyn Myss, Ph.D, who I was reluctantly dragged to see in Austin. Wow, what a dynamic speaker and powerful person. Her work on archetypes helped me accept the teacher in me as well as be open to new ways of thinking and learning through her many insights.

Dr. June Smith was instrumental in assisting my understanding of the Course In Miracles at a time in my life when I was so very open to a spiritual awakening. For Dr. Smith's tireless work and patience every Tuesday for seven years, I am very grateful. Her interpretation of the Course In Miracles helped expand my intellect and let go of made-up inventions around what I was capable of.

Marianne Williamson opened my eyes to being able to embrace the teachings of Jesus and opened my eyes to the power of forgiveness, relationships and love.

And truly one of my most amazing teachers is Vicki Andrews, my friend, spiritual teacher and student of life in a big, big way. Through her unbelievable caring and heart and faith, she helped me discover my soul and the goodness with in. You saw my light long before I knew I had a light!

I also want to acknowledge my partner, Cara Hawthorne, for countless hours of forgiveness, patience and understanding. You opened up a part of me that had been buried under lock and key, and for that I am eternally grateful.

Special thanks to my Energy Advisors at Energy Arts Alliance, J-Coby Wayne and Kain Sanderson, who had faith in me and this book and the journey and purpose I am intended for. They were instrumental in seeing this book through with their valued and selfless hours spent editing, contributing and navigating this ship called *Get Real, People!*

And lastly, my precious Mom, Freda, who reflects all that is good in the world and sometimes blinds me with her light. Posthumously, my thanks to Alan and Normie, my brothers who have gone before me, oh how I miss you both!

Sources of inspiration also include the following authors and bodies of work:

Thich Nhat Hahn
Don Miguel Ruiz
Scott Peck
Gary Zuckoff
Richard Bach
Eckhart Tolle
Krishamurti
Kairos Foundation - More To Life
Landmark Education - Curriculum for Living

## APPENDIX I
## MASTER LIST OF LINKS FROM THE BOOK

www.fb.com/springboardworks
www.google.com/analytics
dictionary.com
www.rolighetsteorin.se
www.viralblog.com/online-video/volkswagens-viral-
video-serie-the-fun-theory/
www.hellomynameisscott.com
www.facebook.com/marismith
www.ceospace.net
www.baxtergraham.com
www.brandyoumadeeasy.com
shelleyroth.com
www.brandyoumadeeasy.com
www.youtube.com/springboardw
www.facebook.com/pages
www.facebook.com/username
www.fb.com/applications
www.fb.com/iwipa
www.fb.com/tabsite
www.fb.com/promotions
www.fb.com/directory
www.fb.com/developers
www.google.com/analytics
www.twentyfeet.com
www.fb.com/sayingitsocial
www.fb.com/baxtergrahamdesignllc
www.fb.com/applications
www.linkedin.com/settings
www.linkedin.com/in/shelleyroth
www.linkedin.com/skills
www.linkedin.com/groups
www.linkedin.com/answers
www.twitter.com

www.shelleyroth.com/store
www.technorati.com
www.alltop.com
www.linkedin.com
www.twitter.com
energyartsalliance.com
www.MarketectsInc.com

**APPENDIX II**
**DETAILED INTERVIEW AND SURVEY RESPONSES**

**INTERVIEW NOTES: MARISOL GRAHAM, PRINCIPAL,**
**BAXTER GRAHAM DESIGN (www.baxtergraham.com**
and **www.brandyoumadeeasy.com)**

> *"As a graphics artist that does web design, branding and logos, I have a unique perspective when I '"read'" companies' websites, brochures, etc. online. As an artist, I can look at a picture and immediately "get" if someone is to be trusted, is filled with light, is dark, etc. The same holds true for a company's "identity" and their digital footprint. The message that company is sending is dependent on so many things artistically that are interpreted by the brain without even thinking. Whether it's the colors they use or the message they have created for their brand, their digital footprint is telling a story about them.*
>
> - Interviewee Marisol Graham

What, in your opinion, does it mean to be REAL?

*To be truthful and honest about who you are, where you come from, what you want to attain, just honest, just be honest as far as being real with someone.*

1.  When you meet someone, how do you know if they are being real? What are some examples? What are some physical cues? Emotional cues? Mental cues? Language cues?

    *At a networking mixer, is their body language too fidgety, not comfortable in their skin, rehearsed spiel in terms of what they say? I try and ask them personal questions so they get out of their mantra... I ask a random question...Where'd you get your belt? Sometimes you never get to see the real person; their eye contact is not focused on you. Instead, they're scanning the room.*

3.  How about online via blogs, Facebook, company website?

    *If some offer sounds too good to be true, it usually is. The ones who choose to be too personal on their Fan Page, that's their business. For example, this woman called out a vendor in front of everyone on her Fan Page. That should have stayed behind the line.*

4.  What concerns did you have prior to using social media tools like Facebook and LinkedIn? (check all that apply):

    ___Privacy

___Security
___Spam
___Not sure how to brand yourself/your company
___Will people understand our brand or
    misinterpret it?
___Other (please elaborate)

*Author's note: No answer given*

5.   Would it be helpful if you could read people's
Digital body language? If yes, can you give
examples of why?

*On your personal page, don't post a picture of you
being a serial killer. Red flag: over-promoting
yourself; communication was overly aggressive,
emails were overly aggressive. Etiquette.*

*Author's note: Didn't really answer the question, but
useful insight.*

5.   What tells you when someone is not being:
REAL/SINCERE/HONEST/FORTHRIGHT/
TRANSPARENT online or on social media platforms?

___Used car salesman syndrome-chest thumper
___Personal Profile Picture-Physical impression
_x_Energy from profile picture-Intuitive feeling
___ Photo albums or videos
___Long, goes on forever, "sales pitch" via email
    etc.
___Buried pricing, have to search high and low to
    find price
___Buried call to action
___Shameless promotion; chest thumping

___It's all about them
___Someone that asks me nothing about myself
___Agenda is apparent
___Hidden agenda-smooth marketer
___Ask for what they want vs. what's in it for me;
    What's in it for them vs. what's in it for me
___ Other. Please elaborate

*I don't like it when people do backhanded compliments. With my profile, I firmly believe I give positivity out there as much as possible. I encourage people with thoughts for the day. I don't like when they do back handed compliments for someone else. "I love how such and such did their website, however, I would have done something different." These people make others feel lesser by trying to elevate themselves.*

6. How do you read someone's Digital body language?

*I will go to someone's page and see how they interact with others on their page. If they don't have good grammar on their page, they're not representing their business well online... You can't misspell words. I also see red flags when words are shortened and there are curse words if you're supposed to be representing your business. For personal use, it's ok to connect to your people by trying to have the hip-hop lingo. You can disrespect others through electronic communication when your communications are short or abrupt.*

7. How do you read a company's Digital body language?

   *Poor design, design changes from page to page…no continuity… That tells me they didn't take time to say how they wanted to represent their company. They pulled some kid to throw something together… They didn't care about their brand. It shows they don't understand how to market themselves, so how long will they last as a company?*

8. How do you know if they are being "real" online (according to your definition of real)?

   *Author's note: No answer given*

9. What are some indicators of "realness?"

   *In various countries colors represent different things. In America, white represents clean, purity whereas in India , white represents death. From a marketing perspective , the only universal color for every country is blue. Depending on the shade, it represents anything from royalty to calmness, tranquility, peacefulness. Inappropriate colors can have a negative effect (e.g., a yoga place that's bright purple and black). You always want to be mindful of your clients and what you want to attract. I brand for what the company wants to do that will represent who they are trying to attract.*

10. What are some indicators of "deception"?

    *Author's note: No answer given*

11. What alarms are set off when you view:

    a. LinkedIn profile page
    b. A Facebook Fan Page
    c. Twitter home page
    d. Blog pop-ups
    e. Website pop-ups

    *If you click the close button and it keeps coming up, I
    don't go back to the page again.*

12. What is something that happened to you online that
    just didn't feel REAL?

    *Author's note: No answer given*

**MARISOL GRAHAM, IN-PERSON BODY LANGUAGE VS. DIGITAL BODY LANGUAGE MAP**

| In-Person Body Language | Digital Equivalent |
|---|---|
| Maintains eye contact | *If someone places a comment under your comment to respond to you* |
| Listens versus talks | *If someone posts something on your wall giving you feedback, kudos or insight, you actually respond to what it is they commented on versus self-promotion; relevancy* |
| Language cues, asks questions | *Ask questions* |
| Arms crossed: frustration | *If you post too many times about how you hate your day, your life,* |
| Texting while talking with you | *It's not exactly real time online, so it's difficult to compare the two.* |
| Someone mirrors or mimics you | *All my Latino friends, I refer to them as "chica", but other people that I am in front of, I respond to them professionally.* |
| Nervous gestures like playing with hair, playing with glasses, squinting—skeptical behavior | *Someone who posts constantly from their cell phone, seriously random things...* |

| Fabricates answers to questions | *You can tell online as well.* |
|---|---|
| Tone of voice | *Hard to get tone across online unless you use emoticons because things can be read in different ways.* |
| Confident | *Topics they talk about, if they are whining, even how they refer to themselves. I had to get over that as well. Directly answers a question, doesn't need constant validation* |
| Approachable | *Profile picture not smiling or laughing* |
| Handshake | *Firm digital handshake = I am so glad we connected, let me know more about your company. Wimpy online handshake = thanks and nothing else is acknowledged about the person or business* |
| Loud, boisterous verbally and non-verbally, pushy | *Someone who posts in all capitals, always having a huge announcement every day like "it's the last day"," it's the last hour", "it's the last minute" and then they start again the next day* |

**INTERVIEW NOTES: RAOUL, PSYCHOLOGIST, BIOCHEMIST, CHEMICAL ENGINEER AND CHEF**

1.  What, in your opinion, does it mean to be REAL?

    *Transparent, to a level that is no fears, no self-serving motives.  My visual picture of transparent being a little bit of a [Star T]rek fan is "a Vulcan mind meld" and truly getting into each other's brains, that is transparent.  In a relationship that means trust with No Shame, No Blame, No Judgment.  That is critical. The need to bond with other individuals is so strongly a part of human nature that we will conform and give what we think others want for acceptance and approval. Performer is meeting some need. Audience could be being manipulated. \*footnote: my belief that we should all be authentic, we should be serving, not self-serving.*

2.  When you meet someone, how do you know if they are being real? What are some examples? What are some physical cues?  Emotional cues?  Mental cues? Language cues?

    *I do accept and I believe that verbal only represents 23% of communication. I am looking for non-verbal. If I want to tell if someone is being authentic, I watch for eye contact, facial muscle shifts (people trained in facial recognition are far more accurate than the polygraph).  I am watching for shifts, watching if they look down, etc.*

3. How about online via blogs, Facebook, company website?

    *I assume that there is an agenda. I assume that someone is trying to take something from me. I have a distrust. People's nature tends to take breaks (falls from grace; we can be very tempted to do things we wouldn't do if we are anonymous) under anonymity. A 280 lb. beer-bellied man can represent himself as 180 lb. perfect Ken doll.*

4. What concerns did you have prior to using social media tools like Facebook and LinkedIn? (check all that apply):

    _x_ Privacy
    _x_ Security
    _x_ Spam
    _x_ Not sure how to brand yourself/your company
    ___ Will people understand our brand or
         misinterpret it?
    ___ Other (please elaborate)

5. Would it be helpful if you could read people's Digital body language? If yes, can you give examples of why?

    *Absolutely it would be helpful. An example of why: Self-preservation of my good name and reputation*

5.  What tells you when someone is not being:
    REAL/SINCERE/HONEST/FORTHRIGHT/
    TRANSPARENT online or on social media platforms?

    *There are inconsistencies. Things that don't fit my
    knowledge base. Examples that don't make sense:
    Baffle them with bullshit... Giving a bunch of
    answers that are designed for purposes of
    camouflage. I am learning to trust myself and if
    something just doesn't feel right and congruent, I
    pay attention to those red flags.*

    *We are so conditioned to believe that if it's in
    writing, it must be right, especially if it's on a
    blog...They take it for absolute truth if read on a
    blog via computer. And we need to begin analyzing
    things on a computer. Even something as widely
    read as Wikipedia. It's just people writing it, and
    there are differences in opinion.*

6.  How do you read someone's Digital body language?

    *There is no such thing as perfection. If it sounds too
    good to be true, it probably is.*

7.  How do you read a company's Digital body
    language?

    *If they are trying to meet all needs, they are
    stretching themselves too thin. I look at a company
    that is good at what they do and isn't trying to be
    everything for everyone. Specializing with a level of
    excellence. Focus.*

8. How do you know if they are being "real" online (according to your definition of real)?

   *It would be telling me what they can do for me and what they can't do for me.*

9. What are some indicators of "realness?"

   *The ultimate solution is too broad for companies.*

10. What are some indicators of "deception"?

    *Old joke: If a women has to tell a man she is a lady, she is not. If a Christian has to tell you they are a Christian, bar the doors. If you have to define yourself with titles versus letting actions speak for themselves, huge red flag. If a company tells me they're the best, greatest, it doesn't mean as much as being targeted on what they do.*

11. What alarms are set off when you view:

    a. LinkedIn profile page

    *If I see someone who is collecting connections versus being purposeful. Frivolous fluff in terms of titles. Promotional products, signature page, shows one-third page of associations, titles. It's all a bunch of fluff.*

    b. A Facebook Fan Page
    *Give to Grow. Somebody who is looking to benefit others, serve others, tells me a lot about the person. Any true person will always share and teach and*

*pass on the knowledge of anything they have, and they can teach others to learn.*

c. Twitter home page
d. Blog pop-ups

*I try to discern opinion from substantiated fact. I use scientific criteria to determine if the reference is based on fact. A good example: I am studying the Old Testament and looking at time span between Joseph and Moses, and there are Egyptologists who discount that Moses and Joseph even existed. And if you look at hard core research, pictures, DNA records etc., you can date the Exodus and Moses and King Ramses the Second. Just because you are an authority, doesn't mean you are correct in your statements. They are negating and not putting forth real evidence. Just because you say it, doesn't make it true. Shows the need for research before believing. Anyone creating a blog should not be insulted by others questioning what they are stating.*

e. Website pop-ups

*See earlier statement on focus and not being a braggart*


12. What is something that happened to you online that just didn't feel REAL?

*Be leery of someone who says I am wrong, I made a mistake. Be leery of anyone who becomes defensive. This is what I use as a screening tool.*

## RAOUL IN-PERSON BODY LANGUAGE VS. DIGITAL BODY LANGUAGE MAP

| In-Person Body Language | Digital Equivalent |
|---|---|
| Maintains eye contact | *Thorough, precise answers* |
| Listens versus talks | *Responding to the real me versus imposing what they want me to hear. Black and white versus color commentary.* |
| Language cues, asks questions | *Reflective listening. Re-verbalizing and stating back to clarify the meaning.* |
| Arms crossed: frustration | *Fanfare and bullsh\*t, no real substance, tells me they are holding me back, apart* |
| Texting while talking with you | *Presentation of someone else's agenda versus listening to your needs* |
| Someone mirrors or mimics you | *You can tell they understand their target audience, if they are mirroring my language online, they understand the audience they are reaching.* |
| Nervous gestures like playing with hair, playing with glasses, squinting— skeptical behavior | *Disorganized websites or blogs all over the place* |

| | |
|---|---|
| Fabricates answers to questions | *Baffle them with BS, they are either answering your question with something that is direct and fits versus a bunch of BS or fluff* |
| Tone of voice | *Bolding, capitalizing* |
| Confident | *Professional impression, colors, humor on a professional level* |
| Approachable | *Ease of use of their tool, blog, website, just flow and be intuitive. Taken the time to construct an organized professional page.* |
| Handshake | *Accessibility. If I have to search to find phone numbers and contact information. That should be readily available. Buried. Hiding out.* |
| Loud, boisterous verbally and non-verbally, pushy | *Author's note: No answer given* |

**INTERVIEW NOTES: BECKY EARL, IT CONSULTANT, DIVERSITY TRAINER AND TEAM-BUILDER**

1. What, in your opinion, does it mean to be REAL?

   *To be clear, concise in ones communications and one's presence. I know people have personas and it's odd, you might think they are something they are not. One single entity that gets presented to the world. No personas if we can help it...Some have them, some don't. Crutch-like role.*

2. When you meet someone, how do you know if they are being real? What are some examples? What are some physical cues? Emotional cues? Mental cues? Language cues?

   *I don't think there are any. There is nothing that lets you see behind the curtain. You make assumptions that people are being real. Total blank, no assumptions, gain those as you spend time. Try not to have a filter (diversity training).*

3. What concerns did you have prior to using social media tools like Facebook and LinkedIn? (check all that apply):

   ___Privacy
   ___Security
   ___Spam
   _x_Not sure how to brand yourself/your company
   ___Will people understand our brand or
       misinterpret it?

___Other (please elaborate)

*Privacy issue is the biggest... Started as true social media, not business... People don't know HR is using this. That's your persona out there...*

4. According to your definition of REAL, how do you know if someone's being "real" online? What are indicators of realness?

*You never know. It depends on the setting, biz to biz communications, then there is an expected environmental understanding....bold language...personal communication, you never know, you have to trust with a skeptical eye.*

5. Can you read a person's Digital body language, and if so, how can you read it?

*I notice grammar, mispellings, punctuation. Examples of how to read: short answers and not a lot of chat indicate directness or someone in a hurry or someone not comfortable chatting. Where you come from in the country may be indicative of how you act and may be misconstrued.*

6. Would it be helpful if you could "read" people's Digital body language? If yes, can you give examples of why?
*Are you kidding? I think you could get insight to the personality, and having insight to the personality, you could prescreen those prejudices and what filter*

*they are using on a lot of things. In business, it would be priceless.*

7. How do you read a company's Digital body language?

   *It depends on the product and who your audience might be. Who do you want to attract? What are you showing.. .Purple doesn't work with nuns. Psychological things about color... Light blue puts people at ease. Colors. Bold vs. pastel, men behind bold, women more softer colors. Size of print, font selected... In the digital world you get a chance to try and put some emotions behind word and text. Unfortunately not a lot of people understand ... If everyone had the same key code, that would be great... No common language that everyone knows about... The font they use tells me something about them... No common language.*

8. What alarms are set off when you view:

   a. LinkedIn profile page

   *n/a*

   b. A Facebook Fan Page

   *Logos, color, font. Notice Fans, crowd mentality says more Fans means more popular. That's my filter. Or more well known. (You can buy Fans note to self.)*

   c. Twitter home page

*n/a*

d. Blog pop-ups

*(Political and religious outta there.) Ads on blogs = paying for space, they are paying for space. They are willing to take the money and they shouldn't do that, it's a personal page.*

e. Website pop-ups

*Author's note: No answer given*

9. What tells you when someone is not being: REAL/SINCERE/HONEST/FORTHRIGHT/ TRANSPARENT online or on social media platforms?

   *I don't know. I make the assumption that I don't know if they are being honest.*

10. What is something that happened to you online that just didn't feel REAL?

   *Haven't had that experience. When I communicate, I try and keep my communications real and keep out of the fray.*

## BECKY EARL IN-PERSON BODY LANGUAGE VS. DIGITAL BODY LANGUAGE MAP

| In-Person Body Language | Digital Equivalent |
| --- | --- |
| Maintains eye contact | *If you put out a question for info. That they answer every single thing you asked them for. They respond directly on subject, no beating around the bush.* |
| Listens versus talks | *They say, "Tell me more." Inquiry vs. advocacy, people can answer a question and advocate their own view. They are advocating for themselves. They are asking a question and then answering themselves.* |
| Language cues, asks questions | *Asks questions online* |
| Arms crossed: frustration | *Maybe stilted answers...short and to the point, non-gentle advocacy* |
| Texting while talking with you | *Author's note: No answer given* |
| Someone mirrors or mimics you | *Author's note: No answer given* |

| | |
|---|---|
| Nervous gestures like playing with hair, playing with glasses, squinting—skeptical behavior | *Author's note: No answer given* |
| Fabricates answers to questions | *?* |
| Tone of voice | *Bold, caps, italics ("just kidding"), underline, didn't mean that. Written word pretty powerful.* |
| Confident | *Someone who is confident is more real. Correspondence is precise, not a lot of fluff. Get a feeling.* |
| Approachable | *Yes, correspondence becomes more relaxed, more personable, less formal.* |
| Handshake | *Author's note: No answer given* |
| Loud, boisterous verbally and non-verbally, pushy | *It's all about them.* |

**INTERVIEW NOTES: DR. JILL CARROLL, ADJUNCT PROFESSOR, RELIGIOUS STUDIES, RICE UNIVERSITY**

1.  What, in your opinion, does it mean to be REAL?

    *To know the truth about yourself, to know what's true about you, for you, and to share that without tricks, strategies, etc. To know the truth and speak it.*

2.  When you meet someone, how do you know if they are being real? What are some examples? What are some physical cues? Emotional cues? Mental cues? Language cues?

    *If they can be with you and maintain eye contact. Are they with you or not? They might be distracted, but I equate being real with being present.*

3.  What concerns did you have prior to using social media tools like Facebook and LinkedIn (check all that apply)?:
    *No concerns about social media. I understand how to protect myself. But though no concerns, certain level of risk you have to take to live in the world and be on the cutting edge of things, so willing to take the risk. My main question was how complicated is this going to be and will it drag me away from my purpose?*

    ___Privacy
    ___Security
    ___Spam

___Not sure how to brand yourself/your company
___Will people understand our brand or misinterpret it?
___Other (please elaborate)

7. According to your definition of REAL, how do you know if someone's being "real" online? What are indicators of realness?

   *I don't think you know immediately, it could be a persona, and in fact, it could be a false persona...Dr. Jill is a persona, it's real, but it's a hat I wear for a role I am playing. I gauge what people say, and assess if it seems snarky and mischief-making ...The blogsphere is filled with hatred, and bitterness and violence. Facebook is more calm, and people block them, and delete what they don't want on there. Pictures don't impact me. You have to interpret people.*

8. Can you read a person's Digital body language, and if so, how can you read it?

   *I might inquire a little deeper. If I don't know who they are, I might click to their profile. Sometimes I have access to their profile...silly stuff, barbs, etc. games, etc. don't impact if they are real.*

9. Would it be helpful if you could "read" people's Digital body language? If yes, can you give examples of why?

*I don't know what digital body language is, but it's not distinguished for me. I just see on the service. You are looking at their words, you don't have body, tone of voice, how they type those words, if they use emoticons, if in caps. If they seem to have online etiquette...preface stuff to get across what you are saying, so people can interpret what they are saying.....twitter, 140 characters.*

7.  How do you read a company's Digital body language?

    *Author's note: No answer given*

8.  What alarms are set off when you view:

    a. LinkedIn profile page

    *Author's note: No answer given*

    b. A Facebook Fan Page

    *None*

    c. Twitter home page

    *Author's note: No answer given*

    d. Blog pop-ups

    *Like to see graphics and pictures of people and what they care about*

    e. Website pop-ups

*Innovation. I want to know why they are innovative, they are thinking...Not static, even in the design.*

9.  What tells you when someone is not being: REAL/SINCERE/HONEST/FORTHRIGHT/ TRANSPARENT online or on social media platforms?

    *Author's note: No answer given*

10. What is something that happened to you online that just didn't feel REAL?

    *If I include all the comments on the [*Houston Chronicle*] blog, at least half of encounters are not real... They are not being authentic, not wanting to engage, wanting to maim..dominate, do violence.... Get 5000 hits per entry...*

**JILL IN-PERSON BODY LANGUAGE VS. DIGITAL BODY LANGUAGE MAP**

| In-Person Body Language | Digital Equivalent |
|---|---|
| Maintains eye contact | *Not disappearing in a common thread or online engagement. If you are leaving, say you are, back to etiquette* |
| Listens versus talks | *Listeners tend to ask a lot of questions. And then wait for an answer instead of asking a question and answering it themselves. You can tell if it's flowing back and forth, give and take, they listened and are now responding to you.* |
| Language cues, asks questions | *Author's note: No answer given* |
| Arms crossed: frustration | *People will often just type things, and then snarl, or a word reflects zzzzz, yawn, typing out your emotions. But if people aren't willing to do that, you don't know. (My eyes are rolled up in my head so far...typing out emotional response.) What % type out emotional response? See it more on twitter than* |

| | |
|---|---|
| | *on Facebook.* |
| Texting while talking with you | *When you are on social media site, you have so many options, you can be multitasking with so many things. It's the nature of the beast. If I want someone's undivided attention, I will ask for a meeting or phone call. I will ask for their attention.* |
| Someone mirrors or mimics you | *High school and college kids are having a hard time understanding plagiarism because everything is copied and pasted and then relinked, etc. The notion of an original IDEA, it doesn't even occur to them, that it would be someone else's idea and credit them for it. As scholars, we gain our original ideas, new contaminations, new arrangements, new slants, new insights.* |
| Nervous gestures like playing with hair, playing with glasses, squinting— skeptical behavior | *When I see comments that people write, misspellings, typos, just using internet shorthand...I wonder, what's the deal? Computer with no* |

| | |
|---|---|
| | *vowels...blatant* |
| Fabricates answers to questions | *Author's note: No answer given* |
| Tone of voice | *You have to be explicit and say what your tone of voice is...This makes me angry...Use punctuation.* |
| Confident | *Pure judgment, the more people write, the less I feel they are confident. If that are trying to explain. They go on and on. I look for laser, be direct, forthright.* |
| Approachable | *I assume the people on Facebook are active, they comment, they post, I assume they are approachable, otherwise, they're not really on Facebook, or they're voyeurs...* |
| Handshake | *Usually, the most times people shake my hand is after a speaking event...The equivalent of that online is a comment after I make a comment, or a blog...They say, interesting, thanks for posting it...That's equivalent.* |
| Loud, boisterous verbally and non-verbally, pushy | *Chest thumpers, you can tell.* |

**INTERVIEW NOTES: ANONYMOUS ARTIST WHO RARELY USES THE COMPUTER**

1.  What, in your opinion, does it mean to be REAL?

    *Doing things without influence of others, societal influences or peer pressure. More easily said than done.*

2.  When you meet someone, how do you know if they are being real? What are some examples? What are some physical cues? Emotional cues? Mental cues? Language cues?

    *I usually don't know when I meet someone. The more I get to know someone; the unreal parts become more obvious as I get to know them. It's not possible for people to keep up that masquerade forever...there are points when the mask falls, and I am very perceptive and see that and have glimmer of the real person. Then that makes the mask that much more obvious when I see the mask again.*

    *Physical is not a deciding factor for me. Some people who are real adhere to social norms about physical appearance and some people, especially in artist world, will wear things that do NOT adhere to social norms for the attention grabbing aspect vs. truly being real. Not adhering to social norms about physical appearance doesn't indicate to me if someone is being real. It's a searching, trying on different ways of being.*

3.  How about online via blogs, Facebook, company website?

*I have low expectations of finding anything real online. "Only believe 50% of what you see and none of what you hear." Quote from a minister when I was growing up. Really stuck with you. Anything's that written is just someone's perception of reality, it's never truly reality. It's gone through at least one filter, and usually several. I lived in Hollywood and to read a review of a party that you went to, you can hardly recognize the hype vs. the actual event.*

*I have seen so little true art on the internet, so often it's a glaring example of bad art. I may be jaded, but so many programmers think they can be artists that they just put an image in as part of the programming of the page or application, but don't understand the artistic aesthetic that could be achieved through an artist's eye. They sell that to businesspeople and it looks better than a blank page so they go with it. Coca-Cola has invested in artists, big advertisers do, their visual image is evident in their advertising. But mostly what I see is cookie cutter templates from basic software. Templified! Art is templified on the net.*

4. What concerns did you have prior to using social media tools like Facebook and LinkedIn? (check all that apply):

\_x\_Privacy
\_x\_Security
\_x\_Spam
\_\_\_\_Not sure how to brand yourself/your company

___Will people understand our brand or
    misinterpret it?
___Other (please elaborate)


5. Would it be helpful if you could read people's
Digital body language? If yes, can you give
examples of why?

*n/a I don't meet people online. If I wanted to meet
people more through internet based social
networking, it might matter. I am more of a face to
face person. All of my social media friends are
people I knew for a long time.*

*I could see the advantage of wanting to learn about
new people online that would be safe and it can be
very overwhelming on number of hits you get, so
how to filter for the type of people you want to
connect with?*

6. What tells you when someone is not being:
REAL/SINCERE/HONEST/FORTHRIGHT/
TRANSPARENT online or on social media platforms?

_x_Used car salesman syndrome-chest thumper
___Personal Profile Picture-Physical impression
___Energy from profile picture-Intuitive feeling
___ Photo albums or videos
___Long, goes on forever, "sales pitch" via email
    etc.
___Buried pricing, have to search high and low to
    find price
_x_Buried call to action
___Shameless promotion; chest thumping

___It's all about them
___Someone that asks me nothing about myself
___Agenda is apparent
___Hidden agenda-smooth marketer
___Ask for what they want vs. what's in it for me;
   What's in it for them vs. what's in it for me
___ Other. Please elaborate

7. How do you read someone's Digital body language?

   *Author's note: No answer given*

8. How do you read a company's Digital body language?

   *Companies I am looking to work for, if they have a cheap looking website, that's a red flag for me. On the other hand a small company that spends too much on a website. My gut tells me you don't care what you are spending on your brand, and so you musn't care about your employees. The $'s a company needs to spend on their brand on the web, it's an expense companies incur now, and some aren't spending $'s embracing the new technology.*

9. How do you know if they are being "real" online (according to your definition of real)?

   *Author's note: No answer given*

10. What are some indicators of "realness?"

    *Author's note: No answer given*

11. What are some indicators of "deception"?

    *Author's note: No answer given*

12. What alarms are set off when you view:

    a. LinkedIn profile page
    b. A Facebook Fan Page
    c. Twitter home page
    d. Blog pop-ups
    e. Website pop-ups

    *Author's note: No answer given*

13. What is something that happened to you online that just didn't feel REAL?

    *Author's note: No answer given*

*Statement/Question: Is there a business world for people who don't use technology? Cart for ice, blacksmith. Community has gone online...*

**INTERVIEW NOTES: RUBY RENSHAW, STATEGIC BUSINESS CONSULTANT**

1.  What, in your opinion, does it mean to be REAL?

    *To share from a place of self-acceptance such that there is no fear of someone judging who you are. To "tell it like it is" right from the beginning, without a hidden agenda.*

2.  When you meet someone, how do you know if they are being real? What are some examples? What are some physical cues? Emotional cues? Mental cues? Language cues?

    *I guess I trust that someone is being real and believe that most people are being as real as they can be in any given situation. In terms of Social Media "meeting" I trust, but it seems like you can get quicker to hidden agenda issues.*

3.  What concerns did you have prior to using social media tools like Facebook and LinkedIn? (check all that apply):

    ___Privacy
    ___Security
    ___Spam
    _x_Not sure how to brand yourself/your company
    _x_Will people understand our brand or
        misinterpret it?
    _x_Other (please elaborate):

*Never sure how much" In-progress "is appropriate
to share (e.g., being real vs. only sharing when
something is solid)*

4.  Would it be helpful if you could read people's
    Digital body language? If yes, can you give
    examples of why?

    *It would be useful from the standpoint of how
    real/serious they are being in terms of working with
    me. I find it hard sometimes to read past people
    being just nice and people seeing something that
    they really want to take on.*

5.  What tells you when someone is not being:
    REAL/SINCERE/HONEST/FORTHRIGHT/
    TRANSPARENT online or on social media platforms?

    ___Used car salesman syndrome-chest thumper
    ___Personal Profile Picture-Physical impression
    _x_Energy from profile picture-Intuitive feeling
    _x_ Photo albums or videos
    ___Long, goes on forever, "sales pitch" via email
       etc.
    ___Buried pricing, have to search high and low to
       find price
    ___Buried call to action
    _x_Shameless promotion; chest thumping
    _x_It's all about them
    ___Someone that asks me nothing about myself
    ___Agenda is apparent
    ___Hidden agenda-smooth marketer
    ___Ask for what they want vs. what's in it for me;
       What's in it for them vs. what's in it for me
    _x_Other. Please elaborate:

*There is a format of marketing out there that seems like a template/insincere (e.g., you keep clicking and clicking to get to "the bottom line" and/or buying something and then being told I can add this and this and this at a discount...Go daddy is notorious for this.*

**INTERVIEW NOTES: KELLIE SCHNEIDER, BUSINESS OWNER**

1.  What, in your opinion, does it mean to be REAL?

    *To be "authentic" – to follow through and do what you say you are going to do.*

2.  When you meet someone, how do you know if they are being real? What are some examples? What are some physical cues? Emotional cues? Mental cues? Language cues?

    *They look you in the eye; they stay focused on the conversation you are having. They repeat what you have said or asked and ask questions back. They offer suggestions and ask how they might be able to help you.*

3.  What concerns did you have prior to using social media tools like Facebook and LinkedIn? (check all that apply):

    _x_ Privacy
    _x_ Security
    _x_ Spam
    _x_ Not sure how to brand yourself/your company
    _x_ Will people understand our brand or
         misinterpret it?
    _x_ Other (please elaborate)

4.  Would it be helpful if you could read people's Digital body language? If yes, can you give examples of why?

*It would be helpful, but I can usually tell from a
response, or lack of response, about a person.
However, giving people the benefit of the doubt,
some have a hard time communicating in words and
do better in person, so I think this option is a good
opportunity to see someone in another light.*

5.  What tells you when someone is not being:
    REAL/SINCERE/HONEST/FORTHRIGHT/
    TRANSPARENT online or on social media platforms?

    _x_ Used car salesman syndrome-chest thumper
    _x_ Personal Profile Picture-Physical impression
    _x_ Energy from profile picture-Intuitive feeling
    _x_ Photo albums or videos
    _x_ Long, goes on forever, "sales pitch" via email
        etc.
    _x_ Buried pricing, have to search high and low to
        find price
    _x_ Buried call to action
    _x_ Shameless promotion; chest thumping
    _x_ It's all about them
    _x_ Someone that asks me nothing about myself
    _x_ Agenda is apparent
    _x_ Hidden agenda-smooth marketer
    _x_ Ask for what they want vs. what's in it for me;
        What's in it for them vs. what's in it for me
    ___ Other. Please elaborate

**INTERVIEW NOTES: SONYA, LANDSCAPE DESIGNER**

1.  What, in your opinion, does it mean to be REAL?

*To be consistent each time you are with someone.
Give your full attention. Avoid distractions. Allow a
person to make contact with you. Don't be closed.
Be truly interested in other people.*

2.  When you meet someone, how do you know if they
    are being real? What are some examples? What are
    some physical cues? Emotional cues? Mental cues?
    Language cues?

    *Make eye contact. Open stance, not folded arms.
    Ask questions about you and listen for the answer.
    Interact.*

3.  What concerns did you have prior to using social
    media tools like Facebook and LinkedIn? (check all
    that apply):

    _x_Privacy
    _x_Security
    _x_Spam
    _x_Not sure how to brand yourself/your company
    ___Will people understand our brand or
       misinterpret it?
    ___Other (please elaborate)

4.  Would it be helpful if you could read people's
    Digital body language? If yes, can you give
    examples of why?

    *Yes. You could determine if you want to spend time
    with them.*

5. What tells you when someone is not being:
REAL/SINCERE/HONEST/FORTHRIGHT/
TRANSPARENT online or on social media platforms?

   _x_ Used car salesman syndrome-chest thumper
   ___Personal Profile Picture-Physical impression
   _x_ Energy from profile picture-Intuitive feeling
   ___ Photo albums or videos
   _x_ Long, goes on forever, "sales pitch" via email
      etc.
   _x_ Buried pricing, have to search high and low to
      find price
   _x_ Buried call to action
   _x_ Shameless promotion; chest thumping
   _x_ It's all about them
   _x_ Someone that asks me nothing about myself
   _x_ Agenda is apparent
   _x_ Hidden agenda-smooth marketer
   _x_ Ask for what they want vs. what's in it for me;
      What's in it for them vs. what's in it for me
   ___ Other. Please elaborate

**INTERVIEW NOTES: LILLY**

1. What, in your opinion, does it mean to be REAL?

   *Real person does not intentionally or consciously represents themselves to be anything they are not.*

2. When you meet someone, how do you know if they are being real? What are some examples? What are some physical cues? Emotional cues? Mental cues? Language cues?

   *Eye contact; fidgeting; evasiveness*

3. What concerns did you have prior to using social media tools like Facebook and LinkedIn? (check all that apply):

   \_\_\_Privacy
   \_x\_Security
   \_x\_Spam
   \_x\_Not sure how to brand yourself/your company
   \_\_\_Will people understand our brand or misinterpret it?
   \_x\_Other (please elaborate):

   *The point of it*

5. Would it be helpful if you could read people's Digital body language? If yes, can you give examples of why?

   *Yes. You could determine if you want to spend time with them.*

5. What tells you when someone is not being:
   REAL/SINCERE/HONEST/FORTHRIGHT/
   TRANSPARENT online or on social media platforms?

   _x_ Used car salesman syndrome-chest thumper
   ___ Personal Profile Picture-Physical impression
   _x_ Energy from profile picture-Intuitive feeling
   ___ Photo albums or videos
   ___ Long, goes on forever, "sales pitch" via email
       etc.
   _x_ Buried pricing, have to search high and low to
       find price
   ___ Buried call to action
   _x_ Shameless promotion; chest thumping
   ___ It's all about them
   ___ Someone that asks me nothing about myself
   ___ Agenda is apparent
   _x_ Hidden agenda-smooth marketer
   ___ Ask for what they want vs. what's in it for me;
       What's in it for them vs. what's in it for me
   ___ Other. Please elaborate

**INTERVIEW NOTES: MARIAN LASALLE, BUSINESS OWNER AND NETWORKING QUEEN**

1. What, in your opinion, does it mean to be REAL?

   *I find that the phrase [Energy Arts Alliance] came up with "Give to Grow" is exactly what I look for in myself and the people I meet. I want a givers gain attitude in the people I do business with and when choosing my close friends. To be _real_ is to care about more than you. Under normal circumstances you can tell fairly fast if the person you're meeting is REAL.*

2. When you meet someone, how do you know if they are being real? What are some examples? What are some physical cues? Emotional cues? Mental cues? Language cues?

   *Language cues are obvious. They start talking about themselves and never ask about you or give you a chance to get a word in the conversation.*

   *Emotional and mental cues are a bit more subtle but they are just as important. When you feel a person is not being sincere you just know you don't want to do business with them or spend time getting to know them.*

3. What concerns did you have prior to using social media tools like Facebook and LinkedIn? (check all that apply):

   ___Privacy

_x_ Security
_x_ Spam
_x_ Not sure how to brand yourself/your company
___Will people understand our brand or
    misinterpret it?
___Other (please elaborate):

4. Would it be helpful if you could read people's
Digital body language? If yes, can you give
examples of why?

    *It would be very helpful to read someone without
    meeting them so you don't waste your time.*

5. What tells you when someone is not being:
REAL/SINCERE/HONEST/FORTHRIGHT/
TRANSPARENT online or on social media platforms?

    _x_ Used car salesman syndrome-chest thumper
    ___Personal Profile Picture-Physical impression
    _x_ Energy from profile picture-Intuitive feeling
    ___ Photo albums or videos
    ___Long, goes on forever, "sales pitch" via email
        etc.
    _x_ Buried pricing, have to search high and low to
        find price
    ___Buried call to action
    _x_ Shameless promotion; chest thumping
    ___It's all about them
    ___Someone that asks me nothing about myself
    ___Agenda is apparent
    _x_ Hidden agenda-smooth marketer
    ___Ask for what they want vs. what's in it for me;
        What's in it for them vs. what's in it for me
    ___ Other. Please elaborate

www.ingramcontent.com/pod-product-compliance
Lightning Source LLC
Chambersburg PA
CBHW060529210326
41519CB00014B/3178